The Tears of the Crocodile

The Tears of the Crocodile

From Rio to Reality in the Developing World

Neil Middleton
Phil O'Keefe
Sam Moyo

LONDON • BOULDER, COLORADO

First published 1993 by Pluto Press
345 Archway Road, London N6 5AA
and 5500 Central Avenue
Boulder, CO 80301, USA

Copyright © 1993 ETC UK

British Library Cataloguing in Publication Data
A catalogue record for this book is available from the British Library

ISBN 0 7453 0764 7 cased
ISBN 0 7453 0765 5 paperback

Library of Congress Cataloging-in-Publication Data

Middleton, Neil.
The Tears of the Crocodile: from Rio to reality in the developing world / Neil Middleton, Phil O'Keefe, Sam Moyo.
240p. 23cm.
Includes bibliographical reference and index.
ISBN 0-7453-0764-7 (cased).–ISBN 0-7453-0765-5 (pbk.)
1. United Nations Conference on Environment and Development (1992: Rio de Janeiro, Brazil) 2. Environmental policy – International cooperation. 3. Environmental law, International. 4. Environmental protection. I. O'Keefe, Philip. II. Moyo, Sam. III. Title.
GE1.M53 1993
363.7'0526–dc20 93–5266
CIP

Designed and Produced for Pluto Press by
Chase Production Services, Chipping Norton, OX7 5QR

Typeset by PCS Mapping & DTP
Printed in the EC by T J Press, Padstow

Contents

ETC UK

This book is an ETC UK project. Founded in the Netherlands in 1974 and now established in India, Kenya and Britain, the ETC exists to encourage and support local initiatives towards sustainable development. It recognises that local knowledge and experience are the building blocks for any developmental activity and that those communities for whom aid projects of any kind are constructed must have substantial influence on their design. Employing people from many and varied backgrounds, the Foundation can offer expertise in sustainable agriculture, agroforestry, energy, water supplies, institutional development and training and extension courses. For further information write to: ETC UK, 39 Norfolk Street, North Shields, Tyne and Wear NE30 1NQ.

Acknowledgements

The authors were given considerable assistance by government agencies in the Netherlands and in Sweden. These agencies are large and complex and it would be invidious to identify particular departments and those within them, but SP, PL and PH know how they helped. Particular thanks are due to Chris Howorth, who offered invaluable suggestions and provided much useful material, and to Ian Cherrett and Ann Walsh of ETC. We thank Koy Thompson and Lloyd Timberlake, both of the International Institute for Environment and Development (IIED), for their assistance. Bryan Marsh's comments on an early draft were invaluable. We are also much indebted to Geraldine Mitchell, who read each chapter as it was written and who gave essential editorial advice. Our thanks are due to Diana Russell for the care and skill with which she prepared the script for press. And, above all, we need to thank Gary Haley for his contribution – for effort beyond the call of duty and quality beyond our expectations. The views and any errors in this book are, however, the authors' own.

About the Authors

Neil Middleton has been a lifelong publisher and frequent author – his books include *The Language of Christian Revolution* and *The Best of I.F. Stone's Weekly*. He has written numerous articles, particularly on development issues, and now lives and writes in Ireland.

Phil O'Keefe is Professor of Environmental Management at the University of Northumbria. He has extensive experience in the field of energy and environment and has worked widely in the developing world, particularly in Southern Africa. His publications include *SADCC Fuelwood Project Development 1985–87: 9 country reports, Resolving the Irresolvable: The Fuelwood Problem in Eastern and Southern Africa, Woody Biomass Supply for Household Consumption, Energy Sector Management Assistance Programme (ESMAP): Evaluation* and *Biomass Assessment in Africa.* He has written numerous papers on energy and related sectors particularly in Africa. He is a director of the ETC UK, a research foundation aimed at supporting local sustainable practices.

Sam Moyo is Head of the Department for Agricultural and Rural Development in the Zimbabwe Institute of Development Studies. He is a founder member and secretary of the Zimbabwe Environmental Research Organization (ZERO), a network which promotes scientific and technical development among regional, national and non-governmental energy organisations and institutions in Southern Africa. His publications include *Rural Development in the Regional Context: South East Nigeria* and *Energy for Rural Development in Zimbabwe.*

List of Acronyms

CARE	Co-operative Agency for Relief Everywhere
CDA	Centre for Self-Managed Development
CFCs	chlorofluorocarbons
CIDA	Canadian International Development Agency
CIIR	Catholic Institute for International Relations
CIS	Commonwealth of Independent States
CPR	common property resource
ECOSOC	Economic and Social Council of the United Nations
EEZ	Economic Exclusion Zone
FAO	Food and Agriculture Organization
FOE	Friends of the Earth
GATT	General Agreement on Tariffs and Trade
GDP	gross domestic product
GESAMP	Group of Experts on the Scientific Aspects of Marine Pollution
GHG	greenhouse gas
GNP	gross national product
G7	Group of Seven (Canada, France, Germany, Italy, Japan, United Kingdom and United States of America)
HCFCs	Hydrochlorofluorocarbons
HYVs	high-yield varieties
IBRD	International Bank for Reconstruction and Development
ICIDI	Independent Commission on International Development Issues
IDA	International Development Association
IFAD	International Fund for Agricultural Development
IIASA	International Institute for Applied Systems Analysis
IIED	International Institute for Environment and Development
IMF	International Monetary Fund
IPCC	Inter-governmental Panel on Climate Change

IPSEP	International Project for Sustainable Energy Paths
ITTO	International Tropical Timber Organization
IUCN	International Union for the Conservation of Nature
KWDP	Kenya Woodfuel Development Programme
LDC	least developed country
LEEC	London Environmental Economics Centre
LEISA	low external input and sustainable agriculture
LWF	Lutheran World Relief
MFA	Multi-fibre Arrangement
MFN	Most Favoured Nation
MNR	Mozambique National Resistance
MTO	Multilateral Trade Organization
MVs	modern varieties
NAFTA	North American Free Trade Association
NATO	North Atlantic Treaty Organization
NGO	non-governmental organisation
NIC	newly industrialised country
ODA	Overseas Development Administration
ODI	Overseas Development Institute
OECD	Organization for Economic Co-operation and Development
PCBs	polychlorinated biphenyls
SADCC	South African Development Co-ordination Conference
SARD	sustainable agriculture and rural development
SEWA	Self-Employed Women's Association
TFAP	Tropical Forestry Action Plan
TNC	trans-national corporation
UNCED	United Nations Conference on Environment and Development
Unicef	United Nations Children's Fund
UNCLOS	United Nations Convention on the Law of the Sea
UNCTAD	United Nations Conference on Trade and Development
UNDP	United Nations Development Programme
UNEP	United Nations Environment Programme
USAID	United States Agency for International Development
VLCC	very large crude carrier
WCED	World Commisssion on Environment and Development
WHO	World Health Organization
WRI	World Resources Institute
WWF	World-Wide Fund for Nature

List of Figures and Tables

Introduction

> It is the wisdom of the crocodiles, that they shed tears when they would devour. Francis Bacon, *Essays*

There can have been few major international conferences heralded by such a bad press as that suffered by the world's largest summit meeting, the United Nations Conference on Environment and Development (UNCED) held in Rio de Janeiro in June 1992. Just two days before it began the *Guardian* ran a front-page story bearing the headline 'Rio Summit Crumbling' and, in the same issue, carried a leading article entitled 'Stumbling Down to Rio' (1 June 1992). Two months earlier the *Observer* had printed a story by its environmental correspondent, Polly Ghazi, headed 'Blueprint to Save Planet Lost on the Road to Rio' (5 April 1992). Most of the broadsheets in Europe and the US took similar lines. The refusal of the European commissioner for the environment, Carlo Ripa di Meana, to attend the summit was widely and sympathetically reported, as if, in view of the failure of the organisers to persuade the most powerful governments to agree on more than the most vestigial of agendas, little else was to be expected. Yet this conference was the first summit meeting to have been held on important environmental issues and one of the few designed, at least in theory, to pay attention to the problems of world poverty.

In addition to the Conventions on biodiversity and on climate, a statement of intent on forests and the 'Rio Declaration on Environment and Development', the planners of the conference worked hard and successfully for the adoption, at Rio, of *Agenda 21*. This is a strategic document of some forty sections which is intended to set the framework for environmental and developmental policies for the whole world as it emerges into the twenty-first century. In-fighting in the preparatory committees, particularly by their US members, had substantially emasculated the text and many of the most important issues were rendered almost null by contradictory or over-cautious conditions. The outcome was an immense document of good intentions, made toothless by the rigid exclusion of timetables, serious financial targets, consideration of the terms of international trade and, above all, the role and unaccountability of the multinationals.

All this was occasion enough for pessimism, but President Bush's refusal to sign the framework Convention on Biodiversity in which he saw, as he so quaintly put it, an 'open cheque' in favour of the poor countries or of the environment (it was not always clear which he meant; his talent for obscurity was remarkable), turned gloomy expectation into depressed certainty. In view of the importance placed on the intransigence of the then US president's stand before the conference, it is worth putting it in context. He made plain that he regarded the economy of the United States as of more moment than some uncertain treaty about biodiversity. What he concealed by his remarks is just what sort of fiscal relationship the USA has with the Third World. The figures given in Table 0.1 illustrate the point; a glance at them shows that in the first two years of Bush's presidency, the USA made well over US$100 billion out of the developing world. What, in effect, Bush was reluctant to disturb is, on the one hand, the substantial support that the poor of the world give to the US economy and, on the other, the emerging changes in North–South relationships as they are expressed in new or revised trading agreements. We should not completely discount the value of President Clinton's change of policy on the treaty in early 1993,[1] but it will become clear just how little it really means.

During and immediately after the Summit jeremiads in the press and posturing by some politicians gave way to reports of the 'I told you so' kind, but many of them missed the most important points. Crispin Tickell, for example, sometime Tory adviser on the environment to Downing Street, wrote, in an evasive little article in the *Observer* (7 June 1992), that the rich countries had set up one camp concerned with the environment, while the poor had set up another to do with development – the two failing to communicate. In this judgement he displayed a dubious even-handedness of a kind which we shall have to consider in other contexts later in this book. Here it is enough to say, as it has been said so often before, that the two 'camps' are inseparable, their relationship is organic. There is

Table 0.1
Net resource transfer to the United States 1980–90 from Latin America, Caribbean and other developing countries (US$ billions)

1980	*1981*	*1982*	*1983*	*1984*	*1985*	*1986*	*1987*	*1988*	*1989*	*1990*
-6.9	-6.9	5.8	27.4	41.9	38.7	44.3	57.4	44.9	36.3	46.3
excluding major oil exporters of Asia and Africa which, if included, would add:										
36.7	24.9	5.8	0.8	5.2	3.1	2.3	7.2	5.7	11.3	14.8

Thus $61 billion in 1990 came 'mainly from countries that are accumulating reserves and other foreign assets in dollar investments that are paid for out of dollar trade surpluses'.

Source: United Nations, *World Economic Survey 1991: Current Trends and Policies in the World Economy,* pp. 71–2, New York 1991.

a sort of priority; without understanding the skewed nature of development and considering other possible political and economic courses, we stand little chance of addressing the unquestionably serious issues of the environment.

The purpose of this book is to examine what led to the summit meeting in Rio, the 'Earth Summit', what it addressed and, even more importantly, what it failed to address and what, in the years following UNCED, the social agenda for those involved in the issues should be. To this end we shall, very roughly, follow the order used by *Agenda 21*. We start from the assumption that, catastrophic as the decline in the purposes of the Summit has been, it has at the least, no matter how inadequately and mistakenly, focused public attention where little was given before and may even have provided a platform from which those committed to the war on poverty may compel recalcitrant Northern governments to take some action.

The Secretary-General of the conference, Maurice Strong, supported by Mostafa Tolba of the United Nations Environment Programme (UNEP), used grand words when, before the event, they described the purpose of the Summit. Both men saw in it one of the last chances to 'save the world' and in doing so neatly inverted Marx's dictum[2] by seeking to interpret that world rather than attempting to change it. In theory UNCED was convened to examine progress in the issues raised by the Brundtland Commission (the World Commission on Environment and Development appointed by the UN Secretary-General) since it presented its findings in 1987 in the report entitled *Our Common Future*,[3] better known as 'the Brundtland Report'.

This report makes the point that so long as there is acute poverty, then there will be environmental degradation. An indication of the scale of that poverty may be gained from World Bank predictions set out in Table 0.2, though we should possibly look at them with a little caution. As it seems that the improvement in those figures assumed to take place by the year 2000 is dependent, to a considerable degree, on the continued insulation of the newly emergent South-East Asian economies from the slump at present engulfing the Northern economies, one can only admire the World Bank's blithe optimism. Even so, for most of the South the outlook is bad and, as the Brundtland Report says without qualification, 'A world in which poverty and inequity are endemic will always be prone to ecological and other crises.'[4]

UNCED has turned aside from this approach and has instead invited the world to find solutions to a number of pressing environmental problems, thus shifting the centre of the debate away from development and towards the environment, or, to express the matter more brutally, away from people and on to things, forces. So far as the convenors of UNCED are concerned, climate change as a consequence of an enhanced greenhouse effect, a dangerously depleted ozone layer, rising, polluted and over-fished seas, increasing deserts, the swift extraction of rarer and rarer resources, the growing international shortage of fresh water, the loss of biodiversity and so on are the real problems facing the world now. That these are issues of huge importance to everyone is not in doubt. But in concentrating on the environment and in inviting the South to help in resolving its difficulties while leaving the problems of poverty and development to 'percolate' through could only have been, the present authors are tempted to think, an extravagant exercise in Northern cynicism. Northern bureaucrats have generously acknowledged that everyone, North and South, has a part to play in the fight for environmental stability; after all, they say, it is a common problem, commonly produced. Like all good partisans, they ignore the awkwardness of reality. To give only one example, it is no part of their case to acknowledge that while the North produces some 90 per cent of all carbon emissions it can only reabsorb 10 per cent of them. But the South produces 10 per cent of the carbon emissions and reabsorbs 90 per cent of them. In short, the developing world, for the first time, is being asked to be an equal partner in a world-wide endeavour precisely because the emphasis has shifted away from the needs of the poor. By advancing an environmental agenda the North has once more concentrated on its own interests and has called them 'globalism'.

Table 0.2
Poverty in the developing world, 1985–2000

Region	*Percentage of population below poverty line*			*Number of poor (millions)*		
	1985	*1990*	*2000*	*1985*	*1990*	*2000*
All developing countries	30.5	29.7	24.1	1,051	1,133	1,107
South Asia	51.8	49.0	36.9	532	562	511
East Asia	13.2	11.3	4.2	182	169	73
Sub-Saharan Africa	47.6	47.8	49.7	184	216	304
Middle East & North Africa	30.6	33.1	30.6	60	73	89
Eastern Europe*	7.1	7.1	5.8	5	5	4
Latin America & Caribbean	22.4	25.5	24.9	87	108	126

Note: The poverty line used here – US$370 annual income per capita in 1985 purchasing power parity dollars – is based on estimates of poverty lines from a number of countries with low average incomes. In 1990 prices, the poverty line would be approximately $420 annual income per capita.
* Does not include the former USSR.

Source: Ravallion, Datt and Chen 1992, quoted in World Bank, *World Development Report 1992*, p. 30, Oxford University Press, New York and Oxford 1992.

Our Common Future makes a number of very important points, but it does so in the framework of a muddled set of ideas. While identifying unequal development as the villain in the environment, it suggests that the revival of growth combined with a change in its quality is the cure. This is an answer which, as we shall argue in this book, simply puts the question back by one stage. The report goes on to call for basic needs to be met, for populations to be stabilised, for resources to be enhanced and conserved and for technology to be reoriented. It also makes a strong case for providing classic economic answers to environmental problems. In failing to examine what caused all these demands to become necessary and in not suggesting mechanisms for their fulfilment, *Our Common Future* leaves us to expect merely a modification of business as before.

Even though UNCED has altered the focus of the discussion, the framework of the Brundtland Report still, in some degree, conditioned its agenda. It is worth comparing the agendas in tabular form since we were once supposed to see the Earth Summit as a response to the challenge of Brundtland (Table 0.3). Setting the agenda out in this way emphasises the differences; for Brundtland it was essential to base itself on the needs of people, while UNCED dealt first in global ecological matters. Where Brundtland, in its confused way, at least set out the political dimensions of the world's problems, UNCED evaded them by relegating them to the secondary status of 'cross-sectoral' issues. Nowhere is this more dramatically illustrated than in the list of agreements, some regional and others global, which already exist and which should have been, but for lack of time, brought to the Summit for ratification. They are all resource or environmental agreements; none are to do with the rights and needs of people. These, if they are considered at all, come last, long after some curious twentieth-century consumerist account of 'nature'.

As a result UNCED has effectively marginalised the case that Northern industry made the environmental mess and that any Southern contribution to global pollution is largely a consequence of the uneven patterns of development forced on the South by Northern finance and protectionism. In the same way it has rendered ineffectual the argument that the sale by Third World states of scarce resources, mainly to service debts but also to obtain enough income to take some steps towards development, is compelled by Northern fiscal and trading practices. No one doubts the importance of global carbon emissions or of the need to preserve the tropical moist forests, but by concentrating on these and in trying to make them fit a Northern catchpenny view of 'development', UNCED has evaded the over-riding issue of equity.

It is hard to believe in the apparent guilelessness of some of the Northern framers of this ill-conceived conference, but we must try.

What they seemed unprepared for was the inevitable response from the Third World: what, in return for help in these matters, is the North prepared to do for the South? This question, focused by the Summit, is not new but it is now being asked in a world remarkably changed since 1987. In the intervening years the Soviet Union has disintegrated and, together with the other members of the Warsaw Pact, its successor states are struggling to find ways of introducing what is sometimes called a 'free-market economy'. Even that political symbol of the 'free- market', the Reagan–Thatcher alliance, has disappeared into a growing US isolationism. Germany has been reunited, with major and still largely unpredictable effects on both its economy and its politics. Some of the repercussions of these are already being felt throughout the rest of Europe and are only too likely to increase.

Table 0.3
Two agendas

Brundtland: 'A Global Agenda for Change'	*UNCED: Determining the Fate of the Earth*
A threatened future	Conventions on climate change
Sustainable development	Forests
International economy	Biodiversity, biotechnology, land resources
Population and human resources	Hazardous wastes
Food security	Toxic chemicals
Energy	Freshwater
Industry	
Urbanisation	Actions for sustainable development into the 21st century
	Environmental awareness
The Commons	Poverty and environmental finance
Conflict – environment and development	Agenda 21 – Cross-sectoral issues
Proposals for institutional and legal change	The Earth Charter

Source: Taken from a seminar paper by Phil O'Keefe, John Kirkby and Chris Howarth, December 1991.

The Balkans in particular and Eastern Europe in general are not only in turmoil, but, together with the Commonwealth of Independent States (CIS), have begun to replace the poor countries of the South in the minds of Northern industrialists, bankers and governments as more rewarding objects of 'aid' and development. Most of the 'centrally planned command economies' had highly developed industrial sectors. Polluted they may be, but as exploitable markets for new technologies and as investments for Western industry, they are very attractive. What were state properties have become assets to be stripped by the new governments anxious to buy their way into advanced capitalism. An early example of this may be seen in the purchase by Fiat of a large interest in the Russian car-manufacturing company, VAZ – the makers of Lada motor cars.[5] Late liberal capitalism itself, despite triumphalist claims, is in disarray. Its populations are increasingly polarised between rapidly growing poor and small but powerful rich populations; its industrial, commercial and financial sectors are enmired in deepening crises; its public services are patently collapsing. Even though the world's wealth is concentrated in Northern hands, it is private wealth and looks out on increasing public squalor.

The Gulf War has been the most obvious and dramatic example of a 'resource war' since the colonising wars of the last century. Not only was the control of Middle Eastern oil at stake, but the control of other vital resources too. For example, it was very much in Turkey's interests for Iraq to be reduced in power because of Turkey's exploitation of the waters of the Euphrates, which happens also to be Iraq's principal source of fresh water; a defeated and weakened Iraq is less able to defend its needs against a strong Turkey allied with the victors. So far as this new world can be seen as an 'order', then the USA has emerged as its policeman – smart bombs are its night-stick.

In trying to blame the South equally for polluting activities and in evading the serious issues of development the North is, once again, particularly by means of UNCED, setting the board in its favour. This is particularly true as, with the rise of new technologies and the increasing sophistication of recycling techniques, many resources from Southern countries begin to decline in importance to the developed world. There is a further problem, in this context, in the loss to the South of the ability, at least sometimes, to play off one superpower against another. In this new world, the bargaining power of Southern governments is increasingly limited as their situations deteriorate yet further:

> That poverty is at the heart of the challenge of international development has always been obvious. What is disturbing is that, except in those Asian countries where the thrust of development has been vigorous and sustained, poverty is increasing rather than receding.[6]

Even well-intentioned liberal Northern non-governmental organisations (NGOs) have tended to obscure the issue. In insisting that less attention be paid to the miseries of the poor world and more to their successes in overcoming adversity, the *soi-disant* 'literature of hope', they have assisted in the further marginalisation of the human rights of the deprived. Stone bunds built in Burkina Faso to conserve both top-soil and water may well be important, but the international financial and political carve-up which leaves that country's inhabitants so lacking in basic human needs is even more so.

It is central to any consideration of the conduct of Northern policy in the South to recognise the increasing tendency to shift policy operations on to NGOs and major financial operations onto para-statal bodies like, for example, the Asian Development Bank, the Commonwealth Development Corporation, the World Bank and the International Monetary Fund. NGOs played an important, though restricted, part in the UNCED conference and that role, together with the kinds of body the UN agreed to recognise as NGOs, must also form part of our examination of the Summit's agenda. We must see where the UN is trying to lead us and to show how those parts of UNCED's programme which were important might be incorporated into another framework, one which must be directed to ending Northern indifference to the needs of the poor and which must be informed by making a priority of international social justice.

Even though, in practice, NGOs are major instruments of policy for Northern states they were seen on the television screens during the conference as part of the troubled 'alternative summit' or 'Global Forum'. It is the nature of contemporary news reporting to frame otherwise complex events in the capsule of a 'story', known in contemporary cant as a 'bite'. No matter how well-intentioned the reporters and television film-makers may be, this commonly results in curiously distorted and, usually, sanitised accounts of events; thus 'news' replaces history.[7] In the case of the NGOs at Rio this phenomenon had two, interestingly related, effects. Reporters frequently sought comments from major NGOs such as Friends of the Earth on the Summit's events of the day, but, in doing so, often pictured them in the context of some of the more arcane and religious groups like those banging drums in support of mother earth. The effect, probably not intended by the reporters, was to hint to their audiences that the NGOs and their comments really belonged to the idealistic margins.

The second effect was, in some instances of reportage, to give another impression. Images of smartly suited and smoothly unreal delegates to the official Summit, chiefly from the United States and Britain, offering, in carefully controlled sentences, uncertain accounts of the positions of their governments were simply depressing when compared with the enthusiasm and directness of the spokespeople from the NGOs. However disregarded the views of the NGOs may have been at the Summit, these images almost certainly increased Northern domestic alienation from the passionate self-interest of their governments.

The rich and powerful will not simply cede their advantages, nor will the *force majeure* of a wounded environment compel change, because catastrophe will be cumulative and strike in little bits. The rich world with its obsessive concern with immigration controls is already reacting to the pressures exerted by an increasingly impoverished South and will dig in yet further. Once again the process will be gradual, but inexorably a fortified and isolated developed world is emerging. It is in this context that we should consider one of Boris Yeltsin's more xenophobic remarks made in a speech delivered to NATO in his application for Russian membership: 'We fully support the efforts to create a new system of security from Vancouver to Vladivostok.' Even more dismaying was the response of NATO's secretary-general, Manfred Wörner, who said that Russia's continuing possession of nuclear weapons could be justified as 'a deterrent in view of the proliferation of nuclear weapons in Third World countries'.[8] It is sometimes tempting to think that Southern pressure may combine with the disaffection of poor Northerners in some great, even revolutionary, moment – but this was the generous illusion of the 1960s.

So there is a dilemma to be faced. It is not difficult to envisage a series of important agreements in which the world's resources are divided up on a more equitable basis and it would be straightforward enough to devise means for their implementation. Indeed the Global Forum went some way towards formulating the sorts of treaty that would produce a more equitable world. But such a programme can only serve as a counsel of perfection; it is not clear that, even as a hidden agenda, campaigning for a whole complex of such agreements could make practical political sense. Instead it is necessary to fight, inch by inch, for every single agreement which goes towards improving social justice, and the fight is against entrenched, labrynthine social, financial and political interests and each battle itself will modify the circumstance of the next. We are in the uncomfortable position of being reasonably aware of what sort of action is needed to begin to right major and obvious wrongs, to move towards some kind of sustainable future for our children and to begin to redress the imbalances of unequal development, knowing perfectly well that we shall be fortunate if we live to see very much of it happening.

Nonetheless this book is addressed to hope on a broad front even when that hope is combined with a curmudgeonly expectation. We shall look at UNCED and its presuppositions, neither of which left much room for happy surprise, and suggest what ought to have been. We do so with no very lively expectation of immediate results, but we can at least hope that the exercise will contribute to the politics of sustainability. Perhaps all is not lost: *Agenda 21* has been offered to the world by the delegates to the Earth Summit, in a flurry of self-congratulation, as a programme for the environment and development in the twenty-first century. Many of its proposals can only be welcomed, but, as we have already observed, it is toothless – had it been otherwise it would not have seen the light of day. This is because decisions about equity and the needs of future generations are the stuff of morality and we are operating in a political world from which morality has been banished. In its place, in these mean-spirited times, we find simple greed masked by the euphemisms of 'management' and 'efficiency'. Nevertheless, governments have committed themselves to support certain generalised positions which may yet be fleshed out in myriad local and national campaigns for justice. In *Agenda 21* governments reluctant to disturb the status quo have possibly offered one hostage to fortune too many. It is there that our scintilla of hope may live.

1

The Road to Rio

The Earth Summit was the culmination of a long process which, in a sense, began with the great Third World independence movements following the Northern War of 1939–45, which, significantly, is more commonly known as World War II. Of course the process really began much further back, with the rise of colonialism and the establishment of the European and, later, the United States' political and commercial empires. There are some interesting variations in that history which still affect our perceptions. Most of the temperate zones, like the North and far South Americas, Australia and New Zealand, were taken over by European settlers, usually following more or less successful genocide, and within a relatively short time became independent. Tropical zones, on the other hand, alternatively seen as gardens of Eden or as 'white men's graves' were principally recruited to empires interested in extracting their resources, human as well as natural. Independence for these colonies came only much later and usually when the mechanisms for exercising control over the riches of the tropics without the need for imperial government were in place. This phase is often called neo-colonialism and has the merit of being far more profitable than the most avid of direct colonisers could imagine. But for our purposes we have to take all that history as read. Vast regions of the world existed principally as client states supplying raw materials, food, cheap manufactured goods and cheap labour for the industrialised North and existed, too, as profitable markets, often dumping grounds, for surplus Northern production. Development was uneven because industry was principally confined to the metropolitan North and the colonised world was an exploitable periphery. Following the 1939–45 war the process of these client states achieving at least notional independence began. In every such case, once the battle to oust the coloniser had succeeded, the battle for social justice, for a decent standard of living in the wake of colonial devastation, began.

Newly independent countries joined the United Nations, an association inevitably dominated by the rich and powerful nations of the world (as an instance of this it is worth noting that only two of its six working languages, Chinese and Arabic, are not European in origin). Nonetheless, no matter how compromised, it is the biggest and most influential forum for international pressure. Progress, no matter how small, in relations between the rich and poor nations of the world has generally come about under its auspices. In 1966 it founded the United Nations Development Programme (UNDP) which, although now rarely in the limelight, is the largest organisation in the world in the business of development co-operation, has huge grant-awarding capacity and is the UN's principal development planning and co-ordinating body.

There are a number of major commissions and conferences which signal the course of UN concern with poverty in Third World or 'Southern' countries. The first of these was the appointment, in 1968, of Lester B. Pearson (who for the preceding five years had been prime minister of Canada) to head a commission to examine overseas aid policies and requirements and to look at ways of improving aid and the systems by which it was administered. In the same year the UN Conference on Trade and Development (UNCTAD) recommended that the economically developed countries should devote 1 per cent of their gross national products (GNP) to aid for the Third World – this percentage was to be made up of both state and private (that is, charitable) funds. From 1961 the UN established ten-year programmes, called 'development decades', both for setting targets and for assessing progress. The second of these, which ran from 1971–80, much influenced by the Pearson Commission, called on donor nations to devote at least 0.7 per cent of their GNPs to aid (in this case only state funds were at issue).

Another important moment in this history was the establishment, in 1977, of the Independent Commission on International Development Issues (ICIDI) under the chairmanship of Willy Brandt, an ex-chancellor of the then Federal Republic of Germany. In 1980 this commission published its report, entitled *North–South: A Programme for Survival*, much better known as the 'Brandt Report'.[1] It made a number of significant proposals: a twenty-year programme of extra aid; an increase to a minimum of 1 per cent of GNP in the national aid budgets of wealthy nations; a new system of international taxation to finance support for the poor; funds to be transferred on preferential terms; countries classified as 'least developed' (LDCs) to be allowed more say in the World Bank and the International Monetary Fund (IMF); the establishment of a World Development Fund with universal membership.

Plans for food security were also proposed in the Brandt Report. So, too, were wide-reaching suggestions for the reform of Northern trade practices so as to allow far greater imports of processed goods from the South and to improve the conditions for marketing them. Brandt also proposed the stabilisation of commodity prices and a system for maintaining them, so that poor countries were not perpetually at the mercy of Northern commodity and share markets. Of immense importance was the report's proposal that the industrially advanced countries should provide funds for financing the development of Southern energy sources. In the interim, oil-producing states should maintain their levels of production while the major consumer countries should do their utmost to meet energy conservation targets so that the energy needs of the poor could be met.

Outside the ranks of the conservative presses there were few who quarrelled with the justice and importance of the Brandt demands. However, Northern governments clearly saw in them an uncomfortable and unacceptable agenda which, if followed, might seriously affect power relationships throughout the world. On the one hand many Northern governments owed their existence at least in part to the gigantic financial interests of the developed world and, on the other, the thinking of the Cold War ruled. Poor countries could not necessarily be relied upon to support the right side. There was virtually no international response to the Brandt Report and, shortly after the publication of a second, back-up paper, *Common Crisis,*[2] the Brandt Commission was disbanded. Once in a while efforts are still made to persuade the developed nations to achieve even the target 0.7 per cent of GNP for aid, but only Scandinavia and the Netherlands have either achieved or come close to the Brandt proposals (see Table 1.1). The industrial states as a whole in 1989, the last year for which we have analysed figures, managed on average a mere 0.32 per cent of GNP, while Britain limped along with 0.31 per cent and the United States of America with a disreputable 0.15 per cent.

Pearson and Brandt both chaired developmental commissions. In 1972 another element had been injected into the debates by the

Table 1.1
Overseas development aid as a percentage of GNP

	1970	*1980*	*1989*
Norway	0.33	0.90	1.04
Sweden	0.41	0.85	0.97
Netherlands	0.60	0.90	0.94
Denmark	0.40	0.72	0.94
Canada	0.41	0.47	0.44
Australia	0.59	0.52	0.38
Japan	0.23	0.27	0.32
United Kingdom	0.42	0.43	0.31
USA	0.31	0.24	0.15
Average for industrial countries	0.33	0.35	0.32

Source: UNDP, *Human Development Report 1992*, p. 53, Oxford University Press, New York and Oxford 1992.

United Nations Conference on the Human Environment. Usually known as the Stockholm Conference, or simply as 'Stockholm '72', it was presided over by Olaf Palme, then prime minister of Sweden and a man widely recognised for his liberal humanitarian views. This conference focused concern about the ever-increasing number of environmental problems and pointed up the close links between poverty and the destruction of the environment. It did this so effectively that it is now rare for either issue to be discussed without reference to the other - this is progress of a sort. Stockholm '72 also led to the foundation of the United Nations Environment Programme (UNEP), an organisation designed to keep a close watch on all major environmental problems and to produce schemes for safeguarding the future of the environment. Up until recently UNEP has been hamstrung by under-funding and the need not to upset the governments that provide such funds as it has. *Agenda 21* calls for its greater funding and organisational strengthening [3] and it is possible that, as the threats to the environment grow and as public concern about them increases, UNEP will be able to work more effectively in the ways envisaged at Stockholm.

The very idea of environmental degradation became common

currency in a different way which was, once more, marked by an important conference. It, too, was held in Stockholm in 1982 and was convened by the Swedish government, who invited the signatories of Stockholm '72 to attend. This conference, on the acidification of the environment, did a huge amount to publicise the origins of and the problems caused by acid rain. It was also in the 1980s that disturbing stories of holes in the ozone layer became widespread.

So the stage was set for the most recent international study – that of the World Commission on Environment and Development, named in short the 'Brundtland Commission' after its president, the prime minister of Norway, Gro Harlem Brundtland. The secretary-general of the United Nations called this commission into being towards the end of 1983; its purpose was to look into the alarming rate at which environmental resources were being consumed, at the levels of their waste, particularly in the cause of 'development', and at the ways in which 'developing' countries were falling further and further behind the industrialised world in their standards of living.[4] Three years later, in the spring of 1987, the Commission published its report under the title of *Our Common Future.* It is perhaps worth noting that the three most important environmental-cum-political documents in modern times, *Silent Spring* by Rachel Carson, *Only One Earth* by Barbara Ward and this report were all written by or under the aegis of women.

In a sense the report was all-embracing. It tackled population and human resources, food security, urbanisation, industry and energy, biological diversity, oceans, war, Antarctica and space. One of its key concepts was that of 'sustainable development', and its formulation by the Commission has had a profound and largely beneficial effect on thinking about the state of the world. But the concept is one which the Brundtland Report both failed sufficiently to clarify and which it finally fudged. This is scarcely surprising since the twenty-two (including the chair and the vice-chair) commissioners, who were politicians, academics, lawyers and bureaucrats of widely differing persuasions, could not be expected to agree on a tough, radical position. Nonetheless, the Commission's own short definition will do: 'Sustainable development is development that meets the needs of the present without compromising the ability of future generations to meet their own needs.'[5]

It is not possible to meet 'the needs of the present' without recognising that first among them are the desperate needs of the world's poor. The poor lack adequate food, housing, health-care, sanitation, education, clothes and work. Disparities between rich and poor, most obvious in the South but entrenched in the North too, have, as it were, been institutionalised in the Northern commitment to 'market forces' and the economics of capital. 'Markets' must be protected by batteries of restrictive trade agreements; they are not free, they are rigged and we shall examine some of the ways in which this is done later in this book. During the years of economic expansion of the 1960s and 1970s a boost was given to the economies of the poor countries. Some South Asian and Latin American countries turned from import substitution to export-led growth with considerable success. But with the uncertainties of the 1980s and the emergence of the Northern recession our markets have battened down the hatches. Since the late 1980s it has made no sense to talk of 'developing' countries – huge parts of the world are now spinning down into national collapse and destruction, involving misery, starvation and death for immense numbers of ordinary people. For less emotive indicators see Table 1.2, but even set out like this we can see not merely the ludicrous disparities, the hopeless distances to be made up, but also the chronicle of a situation rapidly worsening. The Brundtland Commission responded to the beginnings of this by urging the need for 'growth' and only remarked that the poor should be enabled to 'grow' sustainably and at a somewhat faster rate than the rich.[6] In this report, as in so many others (see, for example, *Caring for the Earth: a Strategy for Sustainable Living*, which is the Second World Conservation Strategy Project),[7] the necessity of making a decent standard of living possible for the poor of the world, in the North as well as in the South, is reaffirmed, but the meaning of 'growth' is, for the greater part, left undefined and the means of achievement are scarcely, if ever, seriously considered.

A vital shift of emphasis for the debate at large occurred in an early passage in the Brundtland Report, where the Commission suggested that there is also need for a change in the 'quality' of growth: 'The process of economic development must be more soundly based upon the realities of the stock of capital that sustains it. This is rarely done in either developed or developing countries.'[8] The 'capital' referred to here is that of environmental resources. These sentences are vital in two ways: first, they use the language of the balance sheet to make their point; second, while undoubtedly right in its assertion that neither developed nor developing countries maintain their 'stock' of environmental 'capital', the

Table 1.2

Widening economic gaps between rich and poor

	Percentage of global economic activity					
	Global GNP	*Trade*	*Commercial bank lending*	*Domestic investment*	*Domestic savings*	*Foreign private investment*
1960–1970						
Richest 20%	70.2*	80.8**	72.3**	70.4***	70.4***	73.3¶
Poorest 20%	2.3*	1.3**	0.3**	3.5***	3.5***	3.4¶
Ratio, richest to poorest	30:1	62:1	326:1	20:1	20:1	21:1
1989						
Richest 20%	82.7	81.2	94.6	80.6	80.5	58.4
Poorest 20%	1.4	1.0	0.2	1.3	1.0	2.7
Ratio, richest to poorest	59:1	86:1	485:1	64:1	82:1	21:1

* 1960; ** 1970; *** 1965; ¶ developing countries only.

Source: UNDP, *Human Development Report 1992*, p. 36, Oxford University Press, New York and Oxford 1992.

Commission elides the differences in the causes of this failure.

Conceptual problems arise as soon as accounting, or balance-sheet, language becomes the primary way in which we analyse environmental resources. Whether the resource is clean air, a deposit of bauxite or an area of forest, if it is seen as 'stock' then, no matter how far the stricture not to endanger that stock for future generations may be accepted, the temptation to trade unsustainably in it at moments of economic stress will be overwhelming. This problem is made greater because of the need to put a realistic price on so many such resources simply in order to prevent their unrealistic exploitation. The answer may lie in the ability to incorporate the importance of maintaining ecological and resource-use balance as a necessary physical factor in the fight for human survival into current economic thinking. Such a solution would, however, do little for the gravest of all the problems inherent in thinking of resources as 'stock' or 'environmental capital': both descriptions, in contemporary society, imply trade as the necessary context in which to think of them and, as a corollary, raise the issue of ownership and control. Even before that, however, there is a graver epistemological issue. 'Natural' resources are not really natural;

once we see elements in our environment as 'resources' they become one of the ways in which humanity has interpreted or constructed nature. Then to go on to think of those elements as 'environmental capital' is a further economistic and, therefore, political refinement of the human process of building nature and of 'adding value' to it. Who owns these 'natural resources', this 'capital', is a matter which then moves to the centre of the stage and we shall return to this question in later chapters, particularly in the discussion of land as a resource.

Our Common Future fails to distinguish the participants in the 'market'. 'Market forces' compel manufacturers to produce as cheaply as possible (and, incidentally, to sell at the highest price that the 'market' will bear) rather than as environmentally sustainably as possible. Poor countries are compelled to sell their environmental resources to satisfy the demands of those same 'market forces', at prices dictated not by their real value but by the purchasers who control the markets in the first place. Poor people, wherever they are, are in much the same situation as poor countries – if they can sell their labour at all, they must do so for what they can get rather than for what it is worth and the sale, as a consequence, is frequently unsustainable. The apparent even-handedness of the Commission in accusing all the participants of failing to maintain their stocks of environmental capital[9] wears, on second thoughts, a little thin.

Many valuable points were made in the course of the report and a few examples will serve to illustrate this. While deciding that, in some countries at least, attempts to limit population growth should be made, the report points out that:

> An additional person in an industrial country consumes far more and places far greater pressure on natural resources than an additional person in the Third World. Consumption patterns and preferences are as important as numbers of consumers in the conservation of resources.[10]

In dealing with future food security the Commission, among other things, pointed to the farm subsidies of the North resulting in environmentally damaging over-production, the dumping of food surpluses as 'aid' and the consequent lowering of food prices in the recipient countries to the point where poor farmers can no longer afford to produce.[11] The Commission, in considering urbanisation and the massive increase in the movement of Third World people from rural to urban settlements where, as part of the 'informal sector', they provide many of the city's services, produced the laconic comment: 'Their problem is not so much underemployment as underpayment.'[12]

These examples are not offered in any way as an account of the report's contents, but simply to make the point that whatever reservations one might have about its overall emphasis, it does have its excellences. But the reservations remain, on a number of different counts. There is a real sense that developmental issues, including the essential issues of local participation, have been subsumed under economic and environmental imperatives. This need not in itself be a bad thing, but there are considerable dangers in dealing with sustainable development solely in terms of the techniques which go towards preserving the integrity of, or even enhancing, environmental 'capital'. Techniques, too, are important and good practice in agriculture, industry, fishing and so on are obviously to be encouraged. None of this, however, will make the slightest difference either to environmental destruction or to international human justice until the issue of the ownership of resources is resolved.

It is here that the Brundtland Report is weakest. It calls for new initiatives from governments, the UN and the bilateral aid agencies like the World Bank and the IMF. In the first place it demands that all these parties ensure that the developmental projects they support are environmentally sound.[13] It suggests the installation of programmes for the restoration and improvement of the damaged or threatened ecological bases for development. It calls for the better education and training of the experts needed in developing countries and for the strengthening of the indigenous institutions for this purpose.[14] What it does not do is to question the role of the world's dominant governments and institutions in preserving the conditions in which environmental and developmental problems arise.

Certainly there are calls in the report for greater investment to be financed by new forms of taxation, for positive discrimination in favour of poor countries in the international terms of trade[15] and for sensible international conventions and agreements designed to promote environmental and developmental co-operation.[16] The fact that poor countries are caught in the vice of falling commodity prices on the one hand and the need to service the past loans that they have contracted with governments, bilateral agencies and private financial institutions on the other, is recognised.[17] At the same time the culpability of the lending agencies (see Tables 1.3 and 1.4) in maintaining the levels of debt, of the international bourses, of the great financial and industrial corporations, of the wealthy who own shares in poverty goes largely unmentioned. It may be a welcome sign of changing times that *Agenda 21* has a little more to say about debt.[18]

Our Common Future has set the limits to contemporary thinking about development. The relationship between development and environment has been firmly cemented by the report, but in doing so it has achieved a sleight of hand in making the preservation of the environment the dominant feature of the discussion. Similarly, even while making the point that poor countries are effectively subsidising 'the wealthier importers of their products'[19] it applies the strictures about sound environmental practice to them. In proposing special taxes on deep-sea fishing, seabed mining, offshore oil extraction, satellites and so on to finance change, it ties the issues firmly to a form of the present Northern economic system. So we end with a report which, although bowing in the direction of serious developmental issues, is fundamentally

Table 1.3
World Bank – higher lending, lower transfers, 1985–91 (IBRD and IDA)

Fiscal year	*Gross commitments (US$ billions)*	*Gross disbursements (1991, US$ billions)*	*Net transfers (current US$ billions)*
1984	17.7	16.9	5.2
1985	18.5	16.4	4.9
1986	19.7	15.3	3.1
1987	20.3	17.1	2.6
1988	21.2	16.4	0.7
1989	22.8	15.9	0.2
1990	21.5	18.3	2.4
1991	22.7	16.0	-1.7

Note: disbursements from the IDA Special Fund are included. Data for 1984–9 are for current borrowers; those for 1990 and 1991 for all borrowers.

Source: UNDP, *Human Development Report 1992*, Table 4.5, p. 51, Oxford University Press, New York and Oxford 1992.

environmental and economistic. Social justice has become a matter only to be approached through humanising the workings of the market. Yet those workings are clearly, themselves, dependent on maintaining something approaching the status quo.

Other questions lie behind the constitution of the Earth Summit. Stockholm 1982 publicised the problem of acid precipitation and,

Table 1.4

IMF net transfers to developing countries

IMF financial year	*US$ billions*
1983	7.6
1984	5.6
1985	0.2
1983–85 average	4.5
1986	-4.3
1987	-7.9
1988	-7.8
1989	-7.5
1990	-4.2
1986–90 average	-6.3

Source: UNDP, *Human Development Report 1992*, Table 4.4, p. 51, Oxford University Press, New York and Oxford 1992.

more recently, the pictures of the devastation of large parts of Eastern Europe by industrial pollution have driven the lessons home. For a while disquiet about the depletion of the ozone layer was minimal. Northerners did not worry too much so long as the damage was limited to the Antarctic and Australasia – 'Australians spend too long on the beach anyway'. Now the layer is in tatters over the Arctic and over Northern European and North American latitudes, public fuss has multiplied and aerosol deodorants have gone out of fashion. Global warming, despite determined efforts on the part of nervous or venal politicians to brush its dangers aside, has become a commonplace subject for environmental concern.

Everything from energy generation to transport, from factory farming to waste disposal, has become subject to a new 'green consciousness' throughout the developed world. Up to a point it has even provoked a response from finance and industry whose administrators, discovering that it pays, are now marketing goods with a 'greener' image and even manufacturing goods which are marginally less environmentally destructive. One important 'green' response has been the increasing recognition of the transborder nature of pollution and hence also of possible answers to the problems it poses. Tides, wind currents and water-courses will carry pollution indiscriminately across many borders and much effort will be wasted if, for example, the Netherlands tries to clean up a Rhine which France continues to dirty. It is to be welcomed that these matters and many more like them have become common currency in the industrialised world.

One issue has come almost to symbolise all the others – the destruction of the forests. Public concern about this is directed almost exclusively to the tropical moist forests, the rainforests. Their destruction for the production of questionable grazing land (the hamburger clearances), for making paper, particularly computer paper, and, most of all, for trees to supply the international market in hardwoods has roused huge passions. Most recent estimates for 1981–90 suggest that in tropical South America alone 6,800 hectares of forest perished each year.[20] We can summarise the public view: burning and felling the remaining forests increases global warming, acid precipitation and depletion of the ozone layer, destroys biodiversity, wrecks the lives of forest-dwellers and, is entirely caused by entrepreneurial greed, a greed which, its critics feel, has no business to be satisfied by what should be a 'global common'. This greed is often portrayed, particularly by those anxious to preserve the Northern status quo, as Southern: Southern entrepreneurs, corrupt and uncaring Southern leaders, are responsible for selling 'our global patrimony' for dollars.

Exactly how misguided this view is will become plain in our examination of UNCED's priorities. The need for some response to the tide of Northern concern for the globe dictated the formulation of its agenda. Climate change and energy, biodiversity and forests, marine pollution and over-fishing, desertification, the adoption of *Agenda 21*, a declaration of environmental principles, the establishment of the Sustainable Development Commission and the Earth Council, were the topics around which discussion was arranged. In every case the interests of the powerful Northern governments largely prevailed.

The original purpose of the Earth Summit was to look at the world's progress in meeting the relatively modest demands made by the Brundtland Commission and to find ways of initiating some of its more substantive proposals. That Commission's use of environmental and economic approaches to developmental issues, while in itself perfectly comprehensible, has opened the way for an even more skewed debate. World 'leaders' came to Rio not least because those from the industrialised world are under some pressure from their constituents to do something about the environment. Governments are notoriously short-sighted; they can see no further than the next election (or, in some cases, the next coup). Solutions to environmental problems cost money, generally take longer to implement than the life of the enacting government and may well demand substantial changes in the lifestyles of, at least, the fairly well-heeled. Noises may be made by parliamentarians, but so far as they are concerned, 'Decisive action in the hour of need denotes the hero, but does not succeed'.[21] For them, such a summit is a godsend; it both looks like action and it provides a comfortable forum in which to shift blame.

The Summit's final and much watered-down agenda grew out of Northern environmental worries (modified by some Northern desire not to jeopardise its wealth), no matter how real they are. An unintended consequence of that shift in emphasis by the Brundtland Commission is yet another betrayal of the Third World. The agenda postulates a globalism on which it thrusts the interests of the industrialised world. It essentially confines the political and economic solution of the world's developmental and environmental woes to the very methods which led to them in the first place and tries to deal piecemeal with malpractice. To appropriate Beckett, it is a syringe for the costive rich.[22]

Nowhere is this more obvious than in 'The Rio Declaration on Environment and Development', which begins not, as one might suppose, with any reference to the Brundtland Commission recommendations UNCED was originally convened to consider, but with a reaffirmation of the Declaration of Stockholm '72. Although the Rio Declaration is a sort of summary document, and therefore refreshingly brief, the issues of development are passed over with almost indecent haste in three of its twenty-seven 'principles'. They are worth quoting in full:

> Principle 3:
> The right to development must be fulfilled so as to equitably meet developmental and environmental needs of present and future generations.

Principle 4:
In order to achieve sustainable development, environmental protection shall constitute an integral part of the development process and cannot be considered in isolation from it.

Principle 5:
All States and all people shall co-operate in the essential task of eradicating poverty as an indispensable requirement for sustainable development, in order to decrease the disparities in standards of living and better meet the needs of the majority of the people of the world.

From there to the end of the document, 'sustainable development' is increasingly seen as an environmental issue, and in principle 16 we meet the conditioning clause:

Principle 16:
National authorities should endeavour to promote the internalization of environmental costs and the use of economic instruments, taking into account the approach that the polluter should, in principle, bear the cost of pollution, with due regard to the public interest and without distorting international trade and investment.

The Northern infatuation with the 'market' is, of course, not new – it subsists in the very definition of the mercantile capitalism which arose with the end of the late Middle Ages. Industrialisation, finance capital, transnational corporations, world banking have all modified the meaning of 'market', but a linear development is discernible. A whole edifice of philosophy and economics is both produced by it and buttressing it, so much so that alternative approaches are dismissed as intellectually unsound. Political and military defeats of alternative organisations are invariably paraded as proof of their inferiority. Keynes may briefly modify Pareto, but the deification of 'market forces', their 'efficiency' and 'inevitability', usually wins the day.

At the same time, throughout the history of liberal capitalism the pre-eminent nations have perpetually struggled to make sense of the relationship between the state and the people. As a rule there has been an uneasy recognition that unfettered market forces can only end in misery at least for part, if not the larger part, of the working population: the system is based on winners and losers. While the state is rarely in doubt as to its role in providing the proper circumstances for the efficient operation of the market, it has always been equivocal over the question of social justice in the face of market demands.

The state has customarily been the creation of the rich and powerful, their ambition only slightly curbed by popular protest. With the slow enfranchisement of the lower classes came a shift in the state's position. It increasingly announced itself as the rule-maker, the umpire, the mediator between capital and people, no matter that the contestants were unevenly matched. A myth of even-handedness between business and people became common, together with the observation that it was not the role of the state to look after the indigent. For centuries, with small variations and exceptions, education, health, housing, the relief of poverty were the domain of NGOs – they were called charities. In much of the past charities were certainly a palliative, they made tolerable otherwise appallingly deprived lives, but they were also an instrument of state policy. It was not until the final burgeoning of the European welfare states that NGOs in these fields declined in importance. The point, of course, is that Northern states find in NGOs a very useful tool of policy.

Nowhere is this clearer than in the case of aid to the Third World. Questionable Northern aid policies have been blamed on para-statal, international, somehow self-motivating and independent bodies like the World Bank and the IMF. Such bodies are controlled in their perfection by the immutable and universal laws of the market and so cannot be denied. The member states are clearly powerless. All this, at least, is the implication. Banks and financial para-statal bodies apart, there are the Northern NGOs, many of whose members may well condemn state practices privately, but who, in order to achieve their ends, are forced to work within their bounds. There are many benefits to the state in this arrangement. In the first place NGOs are usually charities whose funds come primarily from donations. Even though it is frequently the case that the state is the largest single donor, it would not be too far-fetched to see funds given for the purposes of Northern aid to Third World countries as a surrogate for taxation.

In addition, by ensuring that these charities operate within the framework of the law, by giving them tax relief on income and the occasional handout, and by, from time to time, consulting them, the state can both use them as an instrument of policy in the Third World and ensure that they do not become too effective a channel for discontent at home. A good example of the state attempting to exercise its censorship in matters of this kind was to be seen in the battle between Oxfam and the British government in 1991, when that charity portrayed apartheid as a policy preventing more equal development in Southern Africa. British charity law forbids registered charities to engage in politics and the state, in the form of the Charity Commissioners, threatened to remove Oxfam's charity status on the ground that the condemnation of apartheid was 'political'. That this kind of state policing exists does not detract either from the goodwill of many NGOs or from the undoubted value of much of what they do. The point has to do with their uneasy place as an integral part of the confrontation between the rich and poor countries of the world. No matter how hard NGOs try, and many do try very hard indeed, they remain Northern, white, privileged and, to some degree, compromised as agents of their states.

UNCED did not neglect the NGOs; quite the contrary, it so welcomed them that it has extended the meaning of the term to include a large number of powerful Northern business interests. There was a moment in the preparatory commissions when a major Northern trade association was to be recognised while a number of grassroots African organisations were to be excluded. That particular insanity was stopped, but it is clear that any programme in which it is necessary to fight that kind of battle is deeply flawed. NGOs, both Northern and Southern, were in Rio de Janeiro in force and much was achieved by their meeting one another, even though some fairly rum groups were among them (for example, UNCED recognised as NGO participants, among some sixty or so groups, the Chemical Manufacturers Association, the American Gas Association, the International Organization of Motor Vehicle Manufacturers, the American Mining Congress and no less than three major pressure groups from the coal industry),[23] but they were not equal partners in the UNCED debates: they were invited there as observers and consultants. We shall look further at the role of NGOs later in this book.

Even the venue of the conference is not without interest. Despite an economic downturn in the 1980s Brazil is one of the richest 'Southern' countries, well on the way to being counted among the world's newly industrialised nations. It has within its borders the extremes of Northern kinds of wealth alongside poverty which matches the worst in the world. Two per cent of its landowners control 57 per cent of the agricultural land and there are 11 million landless, or virtually landless, peasants.[24] There is a burgeoning middle class, but there is also a huge and rapidly growing population of 'discarded people', so many of whom were swept out of the sight of the delegates. By 1988 per capita GNP was running at US$2,160, but for the bottom 40 per cent of households it was only $350.[25] Brazil's slum-dwellers are, in many ways, existing in circumstances similar to those of the poor in Dickens' England. So far as its indigenous forest people are concerned, pretty pictures in the press and on television successfully conceal the realities of genocide. However, its spectacular industrial rise, particularly startling in the years 1965–80, is firmly locked into the patterns of Northern economic interest. Its poor need not be an embarrassment; the North invented that invaluable 'trickle down' theory despite all the evidence, and the state of Brazil as a whole can be tricked out as a paradigm for progress. Carnival, after all, is one of Rio de Janeiro's greatest attractions.

There is a sense in which UNCED can be seen as a skilful exercise in propaganda. It has helped to divert attention from the real nature of the issues; it has gulled some Northern campaigners and observers into a sense that something is being done on the international front, though about quite what may be a little less clear; for politicians it is a surrogate for serious action over a whole range of matters on the home fronts of Northern governments; it can be used to fob off, yet again, the legitimate demands of the South; most importantly it could serve to buttress yet further the entrenched institutions of Northern economic control.

Nonetheless, even though the conference itself to a considerable degree turned out to be the lame thing that so many journalists and observers had predicted, it is more than probable that, like Stockholm '72 or the Brandt Commission, its reverberations will be felt for decades to come. In spite of itself, UNCED has opened a can of worms.

2

Climate and Energy

'The Industrial Revolution marks the most fundamental transformation of human life in the history of the world recorded in written documents.' This is the opening sentence of *Industry and Empire* by E.J. Hobsbawm.[1] Certainly, with ever fewer exceptions, the consciousness of the whole world is formed by the continuing history of industrialisation, urbanisation and the consequent economies. Hobsbawm's book is concerned with the rise of Britain as the world's first industrial power, the British Empire, its decline and Britain's relationship with the Third World. His first sentence now has an additional resonance which, at the time of publication (1968), was probably not anticipated. Global warming, the 'greenhouse effect', first mentioned by scientists as early as 1827 and principally caused by past, continuing and increasing energy demands of industrial and industrialising societies, is unavoidable and no one will be unaffected by it.

Climate is created by many differing factors. To recognise this we have only to think of the ways in which special geographic features, like mountains, produce their own weather or, to use the jargon, their own micro-climates. The radiant energy of the sun is the force which ultimately controls our climate. This energy is reflected or absorbed and re-radiated by the molecules of atmospheric gases, clouds and the surface of the earth. Innumerable interactions in the earth's evolution have produced a balance between the energy entering its atmosphere and the energy leaving it. This process is very nearly reproduced by an ordinary greenhouse, hence the life-supporting atmosphere of the earth is itself effectively a 'greenhouse'. Human, particularly industrial, activity during the last century and a half has by accelerating the process upset this balance. In the cause of accuracy, it is therefore better to think of an 'enhanced greenhouse effect', but this is a cumbersome phrase, so for our purposes we will retain the common usage.

Recent studies have tended to refer to 'climate change' rather than to global warming and, indeed, a recent international group formed to assess the question was called the 'Inter-governmental Panel on Climate Change (IPCC)'. Climates change continually, even cyclically, but very slowly, and it may not be possible simply to say that particular alterations are direct consequences of warming. The contribution to change by the production and use of energy in industry and transport, together with industrial changes in farming practice, will modify, accelerate and increase some alterations which might have happened anyway. Alterations in the world's climate can, up to a point, be predicted. What seems not to be possible – and in one sense it is not even interesting – is to be certain that, at the time it is experienced, any particular phenomenon is directly attributable to the greenhouse effect.

It was this element of uncertainty that, for a while, gave the unscrupulous in Northern governments the opportunity to evade the problem. Margaret Thatcher led the world in demanding more precision before serious action could be contemplated. No matter that scientists everywhere were nearly unanimous in agreeing that Rome was burning, Nero's successors continued to fiddle. But Thatcher, Bush and others discovered, like Lincoln, that all the people could not be fooled, at least not all of the time. In the mind of the Northern public the greenhouse effect was added to acid rain and then to the depletion of the ozone layer to make a totality of a grave, even if poorly understood, threat to humanity's well-being, if not to its existence; the public demand for action grew.

Thatcher's last extensive public response to this was astonishing and it took the form of a speech given in the United Nations General Assembly on 8 November 1989.[2] She nodded in the direction of the dangers of global warming but went on to say that: 'Put in its bluntest form, the main threat to our environment is more and more people and their activities.' Her barely concealed view that the problem lay in the over-breeding South was reinforced when she affirmed the need for 'good husbandry' over 'cut-and-burn' agriculture. So far as global warming is concerned it is 'simplistic' to blame Northern 'modern multi-national industry' because: 'It is industry which will develop safe alternative chemicals for refrigerators and air-conditioning. It is industry which will devise bio-degradable plastics. It is industry which will find the means to treat pollutants and make nuclear waste safe.' She had already remarked that nuclear power, 'despite the attitude of so-called greens – is the most environmentally safe form of energy' and that the only way forward was through economic growth. Such effusions from a leading politician are not to be dismissed lightly, for they are the propaganda of the interests she serves.

In 1988 the World Meteorological Organization in co-operation with UNEP established the IPCC, which presented its report in 1990. Meanwhile, in 1989 the first part of another report, financed by the government of the Netherlands and entitled *Energy Policy in the Greenhouse: From Warming Fate to Warming Limit*, emerged from the International Project for Sustainable Energy Paths (IPSEP) in El Cerrito, California.[3] This report differs from many others in setting out to show what, in the event of any of a number of warming scenarios, needs to be done to avert yet greater problematic changes than those already predicted.

In the report, the project leader Florentin Krause and his co-authors Wilfrid Bach and Jon Koomey take an uncompromising line, though the position from which they start has come to have a familiar ring. During the unimaginable stretches of time chronicled only by fossil data and, more recently, by pollen, seed and other biomass deposits there were many and major fluctuations in the world's climate. But Krause and his co-authors, the IPCC and, come to that, the rest of us are not concerned with this stately progress. Human intervention has induced and is inducing substantial atmospheric changes which will affect temperature and climate over the next century, even over the next few decades. Since the late 1980s there has been a world-wide drought, some of the most devastating results of which may be seen in Somalia, in the South African Development Co-ordination Conference (SADCC) countries, in the forest fires in Indonesia, in the droughts of the western USA and in Australia. The El Niño phenomenon, an occasional dramatic reversal of wind and ocean currents in the equatorial Pacific, seems to be building up again. Floods in Bangladesh may be caused, at least in part, by rising seas. It is possible that any or all of these events are consequences of global warming, but certainty, one way or another, will only belong to future generations.

Scientists seem to agree that doubling the present levels of carbon dioxide or equivalent greenhouse gases (GHGs) in the atmosphere would produce a corresponding increase in mean surface temperatures of somewhere between 1.5° and 4.5° C. Most, however, are assuming a smaller range of between 2° and 4° C.[4] These are mean averages, but, of course, actual temperatures vary from place to place. Krause and his fellow authors offer some general observations: given the present commitment to global warming as a result of carbon emissions so far, seas are likely to rise by 0.5 to 1.5 metres over the next few decades; there will be less snow and fewer glaciers; ocean currents will shift and precipitation patterns will change; weather conditions that we now think of as extreme will become more frequent and there will be more storms, floods, avalanches and major changes in the availability of fresh-water run-off; much soil moisture will be lost through greater evaporation and there will be an increase in heat waves and the length of droughts; there will be less rain in the latitudes which cover North America and Eurasia.[5] Krause points out that some geological evidence suggests that when, for whatever reason, climate is sufficiently disturbed it can very suddenly find a new equilibrium causing abrupt, dramatic and totally unforeseeable changes in the process.[6] The plagues which afflicted the Egypt of the Pharaohs may come to seem small beer.

Measured by our wristwatches, climate, for a complex of reasons, changes slowly, which is partly why the attention span of politicians is so limited. This apparent inertia is matched by another: the slowness with which society will adapt to warnings as opposed to present catastrophe. However, to continue to use up fossil fuels indefinitely in the generation of energy to satisfy worldwide need at the present per capita rate of the developed countries is clearly an impossibility, since most informed estimates agree that to do so would exhaust the remaining deposits sometime around AD2030–2050 and engender a greenhouse effect beyond even the worst of current predictions. Simply to adopt that 'wait and see' policy, so popular among many Northern governments, in which the present rate of increase in carbon and related emissions would continue exponentially until matters become plainer, would result in obviously implausible levels of atmospheric concentration of GHGs within the next century. Fortunately such *après-moi-ism* is decreasing and one consequence, no matter how slight, of a sluggish awakening to reality is UNCED's Convention on Climate.

Krause describes two other possibilities. The first is the best case, advanced by Amery Lovins and his colleagues in 1981, in which the most efficient world-wide use of energy could result, by the year 2100, in insignificant levels of GHG emission. The second is a study put out in the same year by the International Institute of Applied Systems Analysis (IIASA). Its authors assume something between the worst and the best of all cases. They suggest that, in the light of current performance and of industrial and political reactions, cumulative carbon emissions will be some eight times those assumed in the scenario for the best of all possible worlds but still only a quarter of the worst case emissions.[7] The bulk of Krause's report consists of an analysis of the present patterns of consumption and the sorts of international industrial and social agreements and commitments needed to achieve the IIASA levels, themselves fairly alarming, in the relatively near future.

The IPCC had different objectives. It was convened to examine the possible effects of climate change on world agriculture. Martin Parry, professor of environmental management at the University of Birmingham, was the lead author of the report and author of its subsequent expansion to book length.[8] He describes the breakdown of mean global temperatures into substantial local differences; in order to assess these the Panel undertook five regional case studies based on the assumption that, in the near future, little effective action will be taken by governments to limit GHG emissions. These studies demonstrate that, in general, winters will get warmer to a greater degree than summers and that some parts of the world will get drier, while others will be wetter. None of this, expressed like that, sounds very alarming until it is realised just who will be affected and in what ways.

One of the most significant changes could be a reduction in rainfall in some already very stressed regions. Among them are the westerly parts of West Africa, the Maghreb, Western Arabia, the Horn of Africa, Southern Africa, South-East Asia, Central America and eastern Brazil.[9] These are virtually all areas which are even now either finding it difficult to feed their existing populations or are already in the middle of catastrophe. This is largely due to the ravages of the political and economic legacies of colonialism and of past 'aid' in the form of massive and often counter-productive loans, but it is also in part due to climates in which quite small changes can very substantially affect the human and ecological balance. In the developed world Australia is also likely to be drier, but the evidence for this, because of ocean behaviour, is more confused. The grain-bearing plains of the USA, which even now are in trouble with drought, will probably become yet drier. On the other hand the climate in north-western Europe could improve. The net result could well be a world still able to provide enough food for its, by then, much larger population, but in different places, in different ways and at costs which are unpredictable. It would be very rash to assume that world food supplies could remain secure even with the modest climatic changes assumed by the IPCC. At present famines are rarely, if ever, caused by an absolute shortage of food in the afflicted regions, but by the inability of the poor to raise the cash to buy it.

It is in the context of the threats to climates, of the world demand for energy and the technologies and markets designed to satisfy that demand, that we must consider the UNCED Rio Declaration, the Convention on Climate and the notional commitments entered into in the adoption of *Agenda 21*. These documents represent the response of at least the developed world both to the mounting evidence of increased global climatic stress and to the intense desire to safeguard their ways of life. The former accounts for the frequently quite good elements in them, while the latter accounts for their lack of any force.

Chapter 9 of *Agenda 21* , entitled 'Protection of the Atmosphere' distils official UNCED thinking on climate and it proposes four general programmes: the first is to address 'the uncertainties: improving the scientific basis for decision making'; the second is to promote the development of energy technologies, the efficient use of energy and the increased availability of energy to developing countries and to promote more efficient transport, to improve industrial development and to consider the effects of climate on both terrestrial and marine resources; the third programme area is to be concerned with 'preventing stratospheric ozone depletion'; the fourth is to deal with the problems of 'transboundary atmospheric pollution'.[10]

It would be idle to pretend that these are not urgently needed programmes, but it might have been a kindness to the architects of this plan to point out that their roof lacks supporting walls. Of course we need perennially to know more, we need more efficient energy use, less destructive transport, a Third World decently provided with energy and an ozone layer to keep us free of skin cancer. But the first issues to approach are those which condition all these: who owns the energy resources and the advanced technology and how do they distribute them? On what basis are resources and technology to be provided for those outside that magic circle? How do we achieve an adequate level of energy security for its end-users throughout the world without financially yet further shackling the poor? *Agenda 21*, faithfully observing the susceptibilities of the powerful nations, does little more than acknowledge that 'equity' is important.

Paragraph 11 of this chapter on protecting the atmosphere, a passage elaborating the programme area devoted to promoting sustainable development in the context of protecting the atmosphere, is fairly representative:

> The basic and ultimate objective of this programme area is to reduce adverse effects on the atmosphere from the energy sector by promoting policies or programmes, as appropriate, to increase the contribution of environmentally safe and sound and cost effective energy systems, particularly new and renewable ones, through less polluting and more efficient energy production, transmission, distribution and use. This objective should reflect the need for equity, adequate energy supplies and increasing energy consumption in developing countries, and the need to take into consideration the situations of countries that are highly dependent on income generated from the production, processing and export, and/or consumption of fossil fuels and associated energy-intensive products and/or the use of fossil fuels for which countries have serious difficulties in switching to alternatives, and of countries highly vulnerable to adverse effects of climate change.

The succeeding paragraph, devoted to setting out a perfectly sensible programme of action, suggests that most of the proposals in it should be carried out with a particular eye to the developing countries, but it is also careful not to include any discussion of the terms on which any transfer might take place.

This chapter is the first in Section II of *Agenda 21*, which is concerned with 'Conservation and Management of Resources for Development', so perhaps we should not belabour the authors too hard for not covering economics, markets and finance here. Instead we may turn to Section IV, 'Means of Implementation', and in particular to Chapter 33, 'Financial Resources and Mechanisms'; here, at least, in these passages which apply to the whole of *Agenda 21*, we might expect enlightenment. Instead we are offered business much as before. The chapter opens with the suggestion that we should 'identify' new resources to enable everyone, but particularly developing countries, to take effective action to protect the atmosphere. Ways and means of funding those developing countries which would otherwise find it difficult to implement these measures must also be 'identified'. Various mechanisms, including both voluntary and international funds, should be 'considered' to ensure the transfer of environmentally sound technology on a favourable basis.

Because it is obvious that it is in everyone's interest to promote economic growth, social development and the eradication of poverty, money and technology should be provided for the developing countries and this will call for a substantially increased effort from both the international community and developing countries themselves (Chapter 33, paragraphs 2 and 3). As the minatory paragraph 4 proclaims: 'The cost of inaction could outweigh the financial costs of implementing *Agenda 21*. Inaction will narrow the choices of future generations.' This is indisputable, but when it comes to explaining how equity is to be achieved, free trade and access to markets are invoked (paragraph 6); they are to be supported by Overseas Development Administration (ODA) finance (paragraph 16) and funding from development banks, the International Development Association and the Global Environment Facility. Debt relief is mentioned; we are told that it is important to find durable solutions to indebtedness; and in the meantime the members of the Paris Club are urged to follow through their agreement, reached in 1991, to provide some relief for the very poorest countries.

The final proposals are an increase in charitable funding (paragraph 16 [f]) and, very importantly, our old friend the 'Mobilization of higher levels of foreign direct investment and technology transfers should be encouraged through national policies that promote investment and through joint ventures and other modalities' (paragraph 17). Add all these means together, particularly with an airy disregard of world trade patterns, and we seem to be left with the unreconstructed elements that played such a large part in creating the problems in the first place.

When we turn to the Convention on Climate, signed by thirty-six governments, including many of the richest, we need hardly be surprised that in many ways we find much the same story. Its objective is set out in Article 2:

> The ultimate objective of this Convention and any related legal instruments that the Conference of the Parties may adopt is to achieve, in accordance with the relevant provisions of the Convention, stabilisation of greenhouse gas concentrations in the

> atmosphere at a level that would prevent dangerous anthropogenic interference with the climate system.

Early in Article 3, which lists the principles of the Convention, there is a passage which points to the need for attention to be paid to: 'the specific needs and special circumstances of the developing country Parties, especially those that are particularly vulnerable to the adverse effects of climate change.'

It is a little dismaying to realise that both those who drafted this document and those who signed it seemed content with another, somewhat startling, sentence in Article 2 which reads:

> Such a level [of greenhouse gas concentrations] should be achieved within a time frame sufficient to allow ecosystems to adapt naturally to climate change, to ensure that food production is not threatened and to enable economic development to proceed in a sustainable manner.

We think that by 'within a time frame sufficient' they mean 'quickly enough', but what could possibly be meant by the rest of the sentence is not so easy to fathom. No one knows how long it will take 'ecosystems to adapt' or even if they will be able to do so, nor is it known exactly how food production will have to change and over what sort of time. It is clear that all they really mean by this sentence is that a quick response is essential, but it betrays a worrying lack of familiarity with the ideas on which it depends.

Contracts of any kind, and particularly those binding states, are very carefully limited, otherwise the contracting parties may find themselves in all sorts of difficulties. This Convention is no exception. In order to get the thirty-six signatories to agree, let alone other nations which may choose to join later, it was necessary to exclude much that is covered in this chapter. The Convention mainly addresses effects, not causes, and it would be unreasonable to criticise it on that ground. As far as it goes, it is in many ways an excellent agreement and does even hint at one cause:

> The Parties should cooperate to promote a supportive and open international economic system that would lead to sustainable economic growth and development in all Parties, particularly developing country Parties, thus enabling them better to address the problems of climate change. Measures taken to combat climate change, including unilateral ones, should not constitute a means of

> arbitrary or unjustifiable discrimination or a disguised restriction on international trade. (Article 3, paragraph 5)

The Convention includes sensible agreements on the exchange and freedom of information, training and research programmes, education and the means of extending all these to developing countries. It is no small matter to have got so many signatories to commit themselves to so much of such international importance, so it seems like a churlish act to point to some of its more debilitating weaknesses, but we must.

Article 4 contains the items to which the signatories have committed their countries and one of the most important is set out in paragraph 2 (a). In it the developed countries, together with most of Eastern Europe and the CIS, commit themselves, in the cause of mitigating climate change, to limiting their 'anthropogenic emissions of greenhouse gases ... recognizing that the return by the end of the present decade to earlier levels ... would contribute to such [mitigation].' The gases referred to are those other than the chlorofluorocarbons (CFCs) and halons covered by the Montreal Protocol. Paragraph 2(b) calls for each state, six months after the Convention comes into force, to report on its policies to the others. None of this leads the way out of the failure by the most polluting of developed countries to crack down on their dirtiest industries or to rationalise their transport policies. Even modest suggestions of special taxes on the 'polluter pays' principle, for example, are regarded as suspect because of their inflationary tendency, as they would be passed on to the consumer rather than absorbed by the producer. Profits and margins tend to be sacrosanct.

Whether or not global warming and the consequent climate change prove to have played a part in contemporary disasters (and a sensible caution would urge us to assume so) its origins are perfectly plain. Northern industrialisation and the mammoth increase in capital was built on the profligate use of fossil fuels to produce energy in the cheapest possible way and with the least possible regard to human and environmental consequences. One has only to think of Charles Dickens' marvellous description of Coketown in *Hard Times* – 'Gradgrindery' has passed into the language, but so, too, should 'Bounderbyism'. The industrial habits and the pre-eminence of the demands of capital in the nineteenth and early twentieth centuries have become entrenched and lie behind that greatest of all barriers to change – 'Our duty to our shareholders'. Prior to the disintegration of the centralised economies of the Soviet Union and Eastern Europe, the same profligacy obtained there for not dissimilar reasons.

The World Resources Institute offers a table which makes this point with some force. It is a list which gives the percentages and ranking of the countries with the highest greenhouse gas emissions, showing that nearly 46 per cent of these emissions came, in 1989, from the three largest single sources: the USA (18.4 per cent), then the USSR (13.5 per cent) and the EC(14 per cent).[11] In Table 2.1 we reproduce the World Bank figures for carbon emissions, which tell much the same story, and we give them to counter the not infrequent view that, although in the past the North had a poor record in this area, technology is overcoming the problem. It is generally agreed that future control presents no insuperable technical problems; the problems are encountered in industry's refusal to invest in clean processes. Although some global warming is inevitable because of the past, the technology exists to make fossil-fuel-based energy production and use efficient and clean enough to reduce future GHG emissions to substantially less threatening levels. For the longer term future the development of alternative and renewable sources of energy, quite aside from the disastrous nuclear option, is now quite probable.

In the developed countries of the North energy is consumed at an overall average rate of 7 kw per capita and in the developing countries of the South the average rate is only 1.1 kw. Reducing Northern energy consumption to 3 kw per capita is already technologically feasible, and by the year 2060 this level could be achieved throughout the world.[12] If the problems of inequality are to be addressed this will only be possible if national governments and the energy industries in both the North and the South come to recognise that it is no longer viable to base the concept of development on the Northern model which incorporates these unsustainably high levels of energy consumption. Such levels can only be sustained by increasingly reducing the number of people wealthy enough to afford them. Equally the Northern levels of consumption of other goods and of mobility can only be achieved world-wide by universally excluding the poor, in the developed as well as in the developing world.

Sadly we must conclude that the present world political and economic situation does not lead us to suppose that technical and

Table 2.1
Carbon emissions

Country group	Total emissions from fossil fuels and cement manufacture (million tons of carbon) 1965	1989
Low-income (all) of which	203	952
China	131	652
India	46	178
Middle-income (all) of which	373	1,061
lower middle-income	176	478
upper middle-income*	198	583
High-income (all) of which	1,901	2,702
Germany (FDR only)	178	175
Japan	106	284
United Kingdom	171	155
United States	948	1,329

*Includes most of Eastern Europe, the former USSR and some of the poorer parts of Western Europe.

Source: extracted from *World Development Report 1992* World Bank, TableA.9,p. 204, Oxford University Press, New York and Oxford 1992.

managerial solutions to energy problems will willingly be introduced. Energy as a commodity is an integral part of the market system; whether it is energy for industry, for services or for transport it must be produced as cheaply as possible and sold for what the market will bear. Hitherto its use has been encouraged just as the consumption of all other commodities in the Northern world is encouraged, and the most dramatic evidence for this is the explosion of the culture of the motor car; motor transport, most of which is in the North, consumes roughly one-third of the world's oil production.[13] In addition the main source of nitrous oxide emissions is also the motor car, but few governments seem willing to face the cultural and personality crises which might follow any attempt to deal with that problem. Like all other major commodities, the production of energy is largely in the control of the so-called private sector or of state facilities whose operations are geared to the needs of the private sector. Technical fixes, particularly when they are expensive, lack appeal unless they carry a demonstrable market advantage.

In the South things are other, but before we examine this it is worth reminding ourselves that 'South' and 'Third World' are misleadingly simplifying terms which encompass a large number of very different regions and very different countries within those regions. There are about 120 nations which might be so described; seventy of them have fewer than 7 million inhabitants and thirty of those have less than 1 million. Very few of the smaller states have marketable resources like oil. In some recently industrialised countries like South Korea there is a very high level of technical capacity and a corresponding consumption of energy. In South Korea's case this is because its industry is substantially dependent on Japanese expansion, a dependence which could lead to a subsequent economic fragility.

Figure 2.1

Urban air pollution – average concentration of suspended particulate matter, by country income group

Micrograms per cubic metre of air

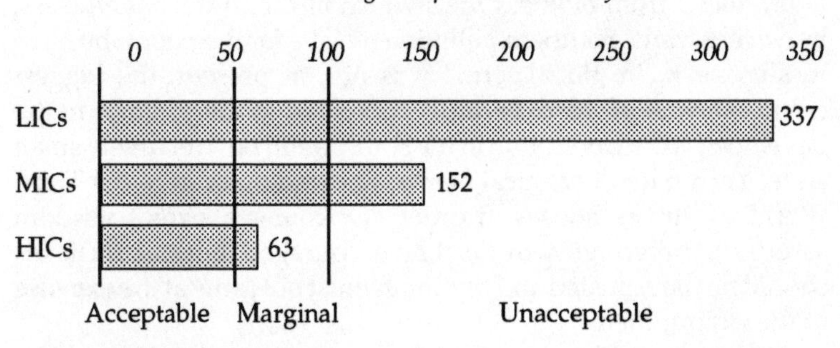

LICs – low-income countries
MICs – middle-income countries
HICs – high-income countries

Notes: a) World Health Organization Guidelines for air quality are used as the criteria for acceptibility.
b) Figures are for 1987–90. Comparable figures for 1979–86: LICs – 323; MICs – 160; HICs – 625.

Source: World Bank, *World Development Report 1992*, Oxford University Press, New York and Oxford 1992, extracted from Table A.5, p. 199 and from Figure 1, p. 5.

In the poorer parts of the South pollution levels caused by industrial and energy facilities and by transport are notoriously high and, at current levels of urbanisation, matters will continue to get worse. It is estimated that by the year 2000 half the world's population will live in urban centres and that the greater part of the change will take place in the Third World. It is also predicted that, by the same date, there will be twenty-one cities throughout the world with populations in excess of 10 million and of these seventeen will be in developing countries.[14] At present, emissions of nitrogen oxides, particulates, sulphur dioxide, carbon dioxide and carbon monoxide from transport, electricity generation, industry and household energy use are creating in the South some of the world's most polluted cities (see Figure 2.1). Not that Northern cities are exactly paradisal; the average figure there gives them only a 'marginal' rating. In general, even in Northern cities, the most polluted areas will be those inhabited by the poorest sectors of the population – the pattern is world-wide. In the South much commercial energy generation is based on coal, which in terms of emissions per unit of energy consumed is the most polluting of fossil fuels. This is particularly so when it is used in poorly designed and out-of-date plants. Some commercial energy and most domestic energy is produced from biomass fuels which also pollute; nonetheless, however serious Southern pollution may be for the poor who have to survive it, in global terms it is not, at present, the biggest environmental problem because, compared to the position in the developed world, pollution in the South is still on a relatively small scale. Yet a patently cynical North is beginning to urge the Third World to put its houses in order. 'Of course', current wisdom asserts, 'Southern growth must be encouraged, but our mistakes should not be repeated and the South must not grow at the expense of the environment.'

For reasons which arise out of a long and familiar history, the South is locked into Northern economic patterns and, in many instances, into specific Northern economies. As it is the South is well aware that it is bound to pay massive tribute to the North in debt-servicing, in providing natural resources and cheap commodities; it also knows that international terms of trade are weighted against it. This becomes painfully clear when we look at its ratios of debt-servicing and cost of energy imports to export earnings (Figure 2.2), while the brute figures for indebtedness and the sources of the loans given in Table 2.2 illustrate the degree of the problem. Most of all, there is a growing and perfectly reasonable dissatisfaction, particularly among the urban poor of the South, with the frequently

Figure 2.2
Energy imports, debt service and export earnings in developing countries

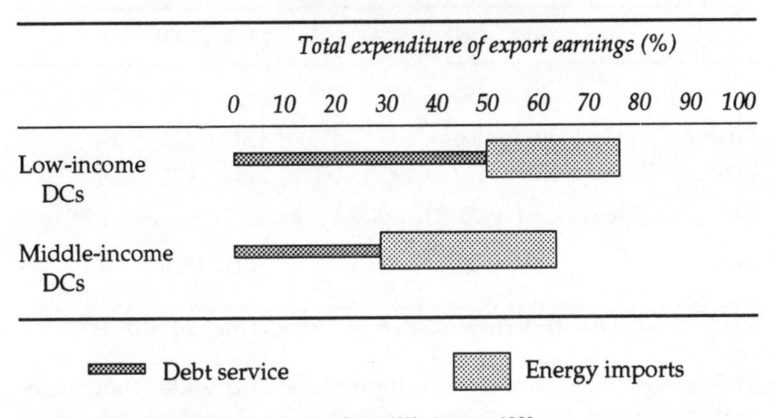

Source: World Bank, *World Development Report*, Washington 1989.

increasing poverty of their lives and the disparities between their conditions and those of their wealthier compatriots and of the citizens of the North. To struggle for growth looks like the only practicable answer and, in a modern and increasingly urbanised world, that growth must be industrial.

Any hope that industrial growth is the solution to the ills of the South may, of course, prove to be illusory; Karl Marx is not the only writer to have pointed out that industrialisation alone did not account for the power of the developed nations.[15] There is the even greater conceptual problem that growth is not only defined in the terms of Northern economic theory (see the Brundtland Report), but is also tied to Northern economies. The primary patterns of trade through which the South hopes to grow lie, as a consequence of financial linkage, between the less developed and the developed world. This is a relationship in which financial control and power are vested in Northern institutions. We must return to the issues of trade agreements and imbalances in a subsequent chapter.

The South also sees the odds stacked against it in other ways. In a competitive world in which the winners have access to the last word in technology (indeed have created it), in which the winners, too, have made knowledge and communication into commodities, the South is largely equipped, when equipped at all, with ageing

Table 2.2
Amounts owed by countries with debt-servicing difficulties (US$ billions).

Years	*1980*	*1982*	*1984*	*1986*	*1988*	*1990*	*1992*
Total	400	550	590	675	665	775	800
Private	300	400	400	415	420	405	390
Official	100	150	190	260	245	370	410
Official as %age of total: 1970–80 declined from c. 35% to barely 30%							
then	29	27	30	40	43	45/48	50

Source: Staff of the IMF, *World Economic Outlook: May 1991*, p. 20, Washington DC 1991.

and worn-out plant and with ill-provided and under-funded research facilities. The frequent Northern habit of sending redundant and obsolete machinery and factories to the Third World and passing them off as aid has not gone unremarked. Above all the question of indebtedness arises. For as long as aid has existed its recipients have been required to pay for it. The price has varied: in some cases political allegiance has proved to be the largest tag, but in most instances political allegiance is combined with healthy interest charges. Aid, too, has commonly been tied to agreements to purchase, with those loans, technology, machinery and expertise from the 'donor' states at prices fixed by those states. In the North deals that look like that would usually be called 'usury' or 'fraud' or, at the very least, 'sharp practice'. We have Ombudspeople and criminal courts to deal with the chancers who set them up.

In a market which, for the countries of the developing world, is anything but 'free', growth has to be achieved against these odds. Economic growth through industrialisation demands an enormous increase in the availability of energy in countries commonly faced with crippling interest charges on loans, including some for the often out-of-date energy facilities that they do possess. Demands that they clean up their act because the North is worried about global warming cannot, in these circumstances, hope to impress.

So far this chapter has referred only to fossil-fuel consumption and its consequent carbon emissions, mainly because in terms of the greenhouse effect, carbon dioxide is the most important gas. It is now common currency that there are other kinds of GHGs, not the least among them methane, generally thought to be, molecule for molecule, twenty times more damaging than carbon dioxide. Of the emissions from anthropogenic sources, rice paddies are estimated to produce about 72 million tonnes per annum and cattle throughout the world vent about 76 million tonnes of it every year; other main human-engendered sources of methane are in the rocketing production of solid waste (46 million tonnes), coal mining (39 million tonnes) and oil and gas production (37 million tonnes).[16] Termites, increasing in number at least in part because of human activity, find tropical grassland particularly attractive and emit surprisingly significant quantities (estimates fluctuate wildly between 5 and 150 million tonnes a year). Methane emissions are increasing at the rate of about 1 per cent a year, and at present are thought to be responsible for 14–18 per cent of the greenhouse effect.[17] The IPCC, thinking solely in terms of stabilising emissions, not of political, economic and developmental realities – which, significantly, were outside its remit – has said that methane emissions should be reduced by between 15 and 20 per cent over the 1990s and the first decade of the twenty-first century.[18] Once again we have an example of Northern concerns making demands which, if pressed, would crucially affect Southern development.

It is in this context that we can consider Krause's principal objective – the rate at, and the means by which, GHG emissions can be reduced to tolerable levels. Apart from carbon dioxide (CO_2) and methane (CH_4), there are three other principal greenhouse gasses: nitrous oxide (N_2O) and the chlorofluorocarbons (CFC-11 and CFC-12). These last two differ from the others because there are no natural sources for them, they are entirely produced by manufacture. Also, unlike the other three, there is no natural 'sink' by which they might, in whole or in part, be absorbed. In early discussions of global warming, risk was estimated in terms of the rates of doubling concentrations of carbon dioxide in the atmosphere. More recent approaches now look at the interactions of all the GHGs and thus at the equivalent of doubling CO_2 concentrations.[19]

The effect of CFCs on the ozone layer and hence on global warming is nowadays at the front of environmental debate. At a meeting of the world's governments in Montreal in 1987, an agreement known as the Montreal Protocol was reached calling for a 50 per cent reduction in their manufacture by the end of the century. Third World countries were to be given ten years longer to allow for the necessary transfer of technology. Further evidence of increased and even more rapid ozone destruction brought about a somewhat less leisurely approach. At a follow-up conference held in London in 1990, it was agreed that the production of CFCs should be entirely phased out by the year 2000, a date which was subsequently amended yet again to 1996. The developed countries also agreed to set up a fund worth US$240 million over three years to go towards the cost of paying for the transfer of technology needed by the less developed countries to abide by these decisions. At the time of writing (April 1993) that fund, not notably generous in the first place, had been reduced to $200 million and its uses were still being determined. Regional workshops were being set up and a newsletter addressed to them was to be launched; criteria for assistance were also to be worked out. Those responsible seem to have an interesting sense of urgency.

That dilatory approach to the fund should have provided us with a warning. Britain, for example, had in 1991 paid over only one-third of its committed contribution and by the end of 1992 had contributed nothing for the year. In November 1992 at a meeting in Copenhagen held under the auspices of the IPCC to follow up the Montreal Protocols, the welcome decision was made to stop the production of, among other dangerous but less damaging chemicals, CFCs by 1996. They will, of course, continue to be released from decaying equipment for some time to come, and as they can stay in the stratosphere for anything up to a century it will be several decades before their destructive effect on the ozone layer declines. Hydrochlorofluorocarbons (HCFCs), marginally less destructive than CFCs and slower to act, were, as agreed at previous meetings, also to be phased out by the year 2005. Pressure from the chemical industry, with British connivance, persuaded the conference to delay the date to 2030. It is widely thought that this decision, next to doing nothing, was the worst of the choices available to the Copenhagen conference. Because HCFCs act more slowly they are likely to coincide with the greatest concentration of CFCs, so the net effect could be that the two chemical attacks on the ozone layer will combine to become the greatest onslaught of them all. Cheap and fairly safe substitutes for CFCs have been produced, particularly in, for example, the German substitution of propane for CFCs in refrigeration units. Made swiftly, this substitution could seriously affect the profitability of the major chemical coporations. The negotiating states clearly felt that such a threat was far more serious than any rapid rise in skin cancers and cataracts, or any damage to land crops and the phytoplankton crucial to the marine food chain, caused by increased ultraviolet radiation. When the pursuit of profit leads to this level of irresponsibility we can only suppose that we are witnessing one of those astonishing acts of *après-moi-ism*. After all, most of the people who agreed to this unprincipled little deal will either be dead, or too old to care by the time its full effects are felt. Brundtland spoke of inter-generational equity, of passing on to our children a liveable world. John Major and his *confrères* have, in this betrayal, expressed their indifference to the demand.

Turning from CFCs to the figures for the world's fossil carbon emissions that were quoted earlier in this chapter, we see that the USA, the states of the former USSR and the EC accounted for 46 per cent of them. Of the remainder China and India account for a further 18 per cent. The rest of the world includes some highly industrialised countries: Japan, Brazil, the newly industrialised countries (NICs) of South-East Asia, Australia, Canada and the whole of Eastern Europe are among them. For many developing countries domestic fuel is overwhelmingly obtained from biomass, principally wood; it is estimated that some 2.6 billion people around the world depend for their supplies on wood fuel, crop residues and dung.[20] Domestic carbon emissions are relatively small and it is in the industries of the developing countries, just as in the North, that we find the main Southern sources of carbon dioxide. However, in 1988 at the Oak Ridge National Laboratory Dr G. Marland and his colleagues produced figures showing that between 1950 and 1986 the per capita production of carbon in the US was twenty-two times that of per capita production of the least developed countries,[21] and in Figure 2.3 we may see the relative rates of energy consumption (the statistics pre-date the collapse of the Soviet Union). It is clear that the environmental priority lies in cleaning up Northern industry and transport.

Markets exist to sell to people who wish to buy and merchants of every kind have age-long and increasingly refined skills in the art of persuading others to buy yet more. At the root of the complexity of contemporary free-market economies is the need to sell and to increase sales in order to grow. Growth, in this view of the world, is a process in which people are enabled to buy more where what is purchased may be seen as enhancing their living standards. There is also a major conceptual problem involved in some of the cruder measures of growth, which often depend on measuring market

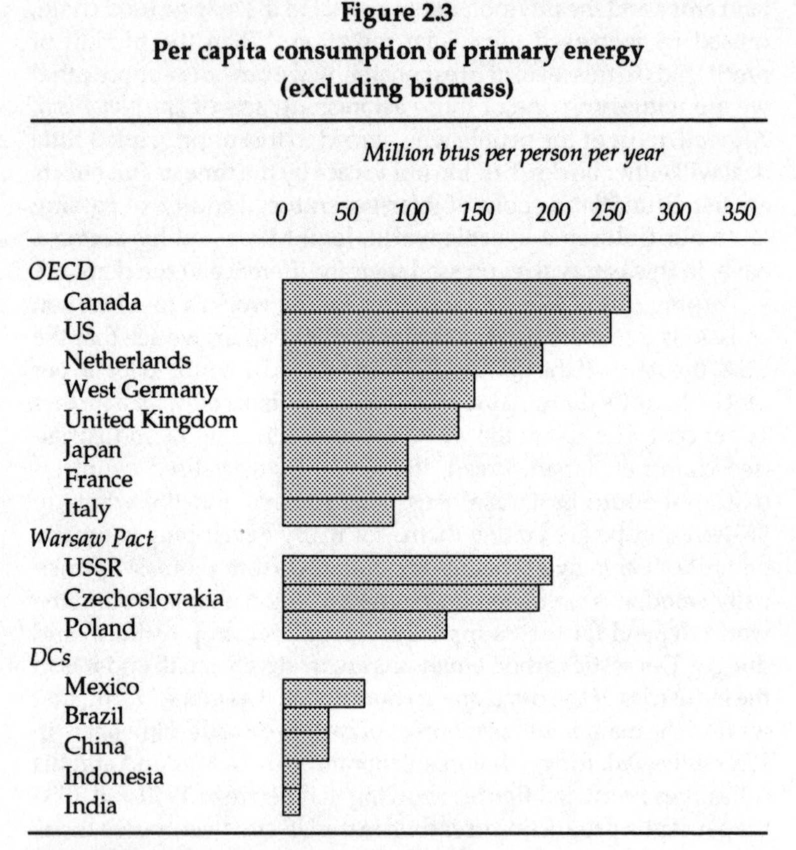

Figure 2.3
Per capita consumption of primary energy (excluding biomass)

Source: US Department of Energy, Washington DC 1989.

performance. It is one which the Brundtland Report has not been entirely successful in avoiding and which is of particular significance in thinking about energy. The market is concerned with the relationship between the volume of sales and the value they realise, but these factors are not enough on which to base a sound developmental and environmental policy. This is simply because market indicators do not take account either of the depletion of natural resources or of the costs of pollution. All that they can do is to discover whether or not energy production is profitable, not whether it is sustainable.

Measures of profitability are always short term; we have only to think of the usual span of those often dubious products called 'business plans' to see this. Because the short-term measure is paramount, markets cannot fail to encourage policies which go towards the lowest possible investment and towards plans which deal only with more or less immediate profitability. Those based not on an arbitrary period determined by rates of return to investors but on the proper life-cycle of any given project will not, as a rule, be considered. As it is improbable that energy can be removed from the commodity market within any foreseeable future, the problems will multiply and if more efficient technology is to become widespread in the South then ways of circumventing the market will have to be found. Otherwise the pattern in which whatever technology is purchased is for the maximisation of profit in the Third World rather than for its efficiency and its cleanliness will be repeated.

To deal with the problems of the greenhouse a major change of direction will be called for. Just as wealth, resource consumption and power are mainly Northern, so, too, is the emission of GHGs. Even if ecologically sound, safe and renewable sources of energy are developed there will still be a substantial transitional time during which fossil fuels will continue to be the primary source of energy. The energy industry must now concentrate on conservation and clean technologies rather than increased sales. Serious energy conservation should, indeed, lead to much reduced sales, which will mean substantial increases in prices. It will also be necessary for the industry and for governments to look at the purposes for which energy is used. Many developed states are actively engaged in encouraging domestic fuel conservation and offering simple technologies to achieve it. In industry, however, including the primary industry of energy production, the picture is quite different. Sectoral interests prevail, with amazingly chaotic consequences.

One of the most obvious areas of absurdity lies in transport; it is common, in the developed economies, for all its forms to be in active competition with one another. In some states, the United Kingdom in particular, there exists what can only be described as an ideological commitment to roads, that most energy-intensive of all transport options. In Britain transport legislation has changed little in character since the 1920s, and at the time of writing the state is continuing to disengage itself from both control and ownership, thus adding to the difficulties of providing the necessary regulation. We may see in *Alice in Wonderland* the most important analysis of British politics to have appeared within the last century and a half.

Transport, with its macho cult of the motor car (and, come to that, the Anglo-French phallic folly, Concorde), may be the most egregious example of wasted energy, but it is not alone. Power stations are notoriously wasteful, even despite the existence for decades of technology which would, to give only one example, allow the heat at present dissipated into the atmosphere through cooling towers, to be used in heating commercial and domestic buildings. Enormous areas of heavy industry, particularly in that section of the newly developing world, Eastern Europe, are based on technologies designed before environmental concerns arose. Yet examples of ways in which energy can be conserved, the ways in which the emission of GHGs can be reduced and that which is emitted can be cleaned up are endless. It is one of many areas of contemporary life where the technological answers are largely available and can be made to work but yet remain unimplemented. The difficulty is both economic and political: economic because Northern energy use is dominated by competing sectoral interests, political because in a 'free-market' economy state intervention is seen as inimical to 'social progress'. Changes in these practices and attitudes will probably only come in response to *force majeure* engendered by a public choking on the consequences of anarchic, free-market energy production.

There is one area in which the North has got its priorities, if not its practices, right: that of energy security. In destroying its coal-mining industry, the British Conservative government is an oddly eccentric exception to this, but then its ideological foolishnesses have always defied analysis. Most Northern states are in no doubt that economic success depends on secure supplies of energy, and a parody of sustainability in ensuring them might almost be detected in the determination of the USA and its allies to secure their future fossil-fuel resources by means of the 1991 war in the Gulf. Be that as it may, security in all forms of fuel is another pre-condition of Southern development and in the proto-developmental stage this must include biomass fuels which are far less readily accounted for in market terms. Except in the form of market research, merchants are unable to assess needs, yet needs must be the first concern in policies designed to secure energy supplies. Markets also have a tendency to centralise and a fundamental condition for successful energy conservation and the creation of energy security will be local, frequently participatory, management for the development of renewable energy resources.

UNCED may be a product of Northern navel-gazing, but even if its priorities are irrelevant to the *favelas*, it at least provides a context for intelligent *Northern* agreements about GHG emissions. Some Northern States – not the least the Netherlands – have already begun to adopt more sustainable policies which might serve as the basis for international and regional goals.[22] Energy policies defined not in terms of sales and consumption but by the real needs of the users would normally lead both to greater energy efficiency, itself an environmental goal, and eventually to the design of technologies which will not suffocate the users and which are based on renewable resources. It would be a foolhardy energy salesperson who would deny that efficient energy use and production make both environmental and economic sense – even the accountants agree that it is cheaper to save an unit of energy than to make a new one – it is merely less profitable. That there is general agreement that energy efficiency is the first and most important step to a rational and safe energy future is indeed a hopeful sign. What holds up the implementation of consequent policies is the momentum of the industry and of its related and dependent services, combined with some of the lunacies we have already described.

Out-of-date and inefficient power stations are still being constructed, motor cars are still rolling off the assembly lines (though in 1992, manufacturers were finding it increasingly difficult to sell them), energy-intensive industrial plant is still being employed with all the abandon of a *wunderknabe* in a spring meadow. All this is simply because producers are fearful that drastic change could compromise their position in the market. Nevertheless, despite a temporary dip in oil prices occasioned by opening the North Sea oil fields and by the politics of the Middle East, the days of cheap fossil fuels are largely over. Energy will, in the future, be a great deal more expensive and this may provide the climate for switching to more efficient use. So far as the urbanites of the North are concerned, cost may combine with cancer and bronchitis to compel a change. One straw in the wind is to be found in a recent British Industrial Survey which, contradicting commonly held managerial views, remarked that rising costs in energy and in materials were of far greater significance than any cost of labour.

Dealing with the vast problems of GHGs, of the conservation of energy resources, of developing other and renewable sources of energy, of ensuring a sustainable future for energy production are parts of a Northern agenda which constantly impinge on Southern priorities. It is sometimes suggested that the developing nations in their route to industrialisation could avoid the mistakes made in the past by the North; that they could 'leap-frog' into modern and sustainable industrial practice. The implication is that the bent back in this game of leap-frog is that of the North. This is an admirable, though possibly wishful, thought, as so far the North has only bent its back enough to pick up the coins in the gutter.

After debt-servicing, the single largest element in the budgets of developing countries is expenditure on energy; the bulk of their balance of payments deficits is accounted for by oil (see Figure 2.2). In order to provide minimally adequate electricity for developing countries the World Bank has estimated that investment of US$125 billion is needed – roughly twice the amount of current investment. In developing countries some 50 per cent of what is invested at present goes in foreign exchange,[23] the economies are small, currencies are weak, the ability to trade internationally is thus circumscribed; the resource base is often narrow and frequently alienated to foreign control, and the burden of debt to Northern agencies, both para-statal and private, is overwhelming (see Table 2.2). These are the circumstances in which the game of leap-frog is to be played, yet it is essential for poor countries to develop and to do so in ways which are less energy-intensive and polluting than those of Northern industry. Technologies which will not further destroy the environment must be made available to them. For this to happen there are certain pre-conditions.

Profitable sections of Third World industry (much of which has been resource-mining) and large-scale agriculture are commonly dominated by transnational corporations ultimately interested in making returns in 'hard' currencies. Dividends payable in Kenyan shillings would lack popularity on the sun-terraces of California (much is to be said for the robust manner of the Cubans in taking over the United Fruit Company shortly after the overthrow of Batista). Whatever facilities for future energy production come to be built in developing countries, it is certain that investment in and control of them must remain in the hands of those countries – they must not become tradeable profit-centres in the international stock-markets. What are we to make of a recent British television advertisement put out by the oil giant, BP? In it an African teenager is shown studying by the light of solar-powered domestic electricity from a unit built and installed by that company.

Most of the principal fuel technologies are to do with the consumption of hydro-carbons and hence are damaging to the environment. For the foreseeable future, much Third World energy will come from their existing older and cheaper forms of production. But many parts of the South need new plant now and for this the cleanest and most efficient technologies must be made available in ways which do not further cripple their economies. For as long as it is necessary for developing countries to import technologies and raw materials for the production of energy, there can be no question of creating international environmental agreements which will penalise the poor for their inability to transform their energy industries in line with Northern programmes.

Caring for the Earth was published by two very large and powerful northern NGOs, the World-Wide Fund for Nature (WWF) and the International Union for the Conservation of Nature (IUCN), in association with UNEP and it is described as a strategy for development. Published a little more than ten years after their first joint document, called the *World Conservation Strategy*, it is an odd document not least because its recommendations are couched in curiously voluntarist language reminiscent of nineteenth-century philanthropy. The opening paragraph of the foreword sets the tone:

> This strategy is founded on the conviction that people can alter their behaviour when they see that it will make things better, and can work together when they need to. It is aimed at change because values, economies and societies different from most that prevail today are needed if we are to care for the Earth and build a better quality of life for all. [24]

The judgement contained in the second sentence is undeniable and much of what is proposed in this document must lie at the base of any sustainable future, but voluntarism is not enough. *Caring for the Earth* by relying on moral urging ducks the central issues of implementation. Nowhere is this clearer than in its recommendations on energy, which in themselves are admirable. Setting the discussion in the framework of the polluting capacity of existing energy production, the document makes clear the importance of analysing needs for energy and the ways of fulfilling these most sustainably rather than most profitably and it sets out a fairly good programme for doing so.[25]

Two examples of the difficulties in the way of such voluntarist proposals will serve. One-fifth of the world's people live in China, which also controls one-third of the world's known coal deposits. It cannot be an easy matter to persuade the Chinese that in the interests of a world which has been less than completely supportive of their enterprises, they should not exploit their reserves in their progress towards industrial development. Saudi Arabia is not a poor country, but it, too, will be dubious about attempts to replace oil with alternative sources of energy. We may hope that Britain is unique in having made the political choice not to have an energy policy and closing down its government Department of Energy.

So the lack of the force of any specificity in the central commitment within the Convention on Climate sits alongside the voluntarism of documents like *Caring for the Earth* and business is much as before. Each government must be pressured by its environmentally aware electorates into giving the Convention and *Agenda 21* some content. Even that battle may be some way off – the Convention has yet to be ratified by the parliaments of the signatory states. Despite the unusual speed with which UNCED actually achieved a Convention, this could yet lead to a measure of uncertainty about the meaning of the words 'by the end of the present decade'.

3

Biodiversity and Biotechnology

Well before the Earth Summit met in June 1992, its proposed Convention on Biodiversity had attracted huge publicity. It was the rock on which the US president's attendance almost foundered and so came very close to finishing off the conference before it even began. Conserving biodiversity had become associated in the public mind with conserving the rainforests, but in the opaque workings of White House consciousness it was interpreted as an assault on standards of living in the US. Neither of these views had much to do either with the reality of the forests or with the subject of the Convention, but there are ways in which both of them are informed by a set of myths which we need to bear in mind.

Dreams of the wonders of the wild were woven by the nineteenth-century romantics. Some, like Wordsworth, organised them and made them urban; others, like Caspar David Friedrich, saw in them the domain of the Nietzschean overman; for Conrad they held the heart of darkness. Thoreau, the ultimate anthropomorphist, simplified matters by making nature uniquely the province in which the individual soul transcends vulgar society, 'All nature is your congratulation',[1] and thus gave rise to hippiedom. The culture was and is magnificent, but, because it was the culture of urbanisation and of capital, it also served as the propaganda for mercantile and imperial expansion. It completed the transformation of the hieratic class relations of the Middle Ages into the power relations of property, for which Luther was the prophet, and so paved the way for Marx's class analysis. Wilderness was wonderful not just because of its fabulous beauty (which could be rendered safe in the suburban neatnesses of, for instance, Mendelssohn's overture, *The Hebrides*[2]) but also because it was a repository of untold 'natural wealth' available only to the resolute. Mining the 'wild' for its human resources in the form of slaves had become unprofitable early on and so was abandoned, with much pious exhortation to the lately enslaved to be good in their new-found 'liberty', in favour of lesser forms of vassalage *in situ*. But the extraction of every other kind of resource was seen as the right of civilised conquest. El Dorado was the name of the game. It has taken the greater part of another century for the recognition of what that romantic propaganda concealed to dawn. Contemporary patterns of pollution are mainly the direct consequences of that colonialism and are, therefore, among the grossest marks of Northern racism.

Romanticism was, in a sense, the ideology of conquest simply because it processed the urban experience of nature into the stuff of myth, indeed made of its objects new myths. In doing so it took humanity out of nature and turned men (it largely excluded women) into supra-natural observers, it turned us from husbandry to mastery. At its centre lay the most potent idea of them all, that of progress, the myth in which the wild, nature, was there for 'man' to tame. Human nature, too, come to that: Gradgrind[3] was the caricature of a common enough figure. The exaltation of the businessman into shaman and moral arbiter allowed the real world, altogether a baser stuff, to be dealt with in a fairly practical fashion: dug up, chopped down, processed on the spot or carted away. It scarcely mattered; this was what God and the Empire had sent 'men' to do and nature was pretty endless anyway. Consider the complex and depressing figure of Cecil Rhodes striding into Southern Africa for the sake of the British Empire, which he once claimed was his only religion.[4]

'Mining' the wild and pressing its 'products' into the service of that Darwinian pinnacle of excellence, 'Western civilisation', has been so extensive that urban, and particularly Northern urban, humanity is now in the position of the child who pulled its toys apart to see how they worked. We have dumped wastes in the seas so that mangrove swamps not cleared by agricultural or urban colonisation are dying and coral reefs are disappearing. These are the two greatest sources of diversity among marine species. We have also fished recklessly so that the seas are emptying; riverine ecologies have been destroyed the world over by insensitive damming and by pollution; above all we have cut down all but the great tropical moist forests and are now well on the way to despatching those too. It is slowly being borne in on us just what it is that has been destroyed and, as a result, why it is so important to save what remains.

Biodiversity is a neologism creeping into common language and is used to cover both species and genetic diversity. Species diversity refers, obviously enough, to the differences between species. When, in the eighteenth century, Carolus Linnaeus invented binomial nomenclature, naming the animals and plants must have seemed an encompassable proposition. There have proved, however, to be rather 'more things in heaven and earth' than we had supposed: estimates now shoot between 5 million and 30 million species, of which only 1.7 million have been studied and classified.[5] A huge proportion of the world's species are insects, but there are also untold numbers of as yet little known plants. Relatively few species, either of animals or of plants, have been domesticated. Most of our agriculture is based on four groups of animals and three grasses, two pulses and one tuber, though with the continued increase in the world's population more may yet be incorporated into the human economy.

Genetic diversity refers to differences within species and is, in a sense, what all the fuss is about. Occasionally we discover that the phrase 'genetic resource' is used indiscriminately to mean both species and genetic diversity, but it really should only refer to those animal and plant characteristics which either might be, or are known to be, of use to humanity. Given that knowledge of species is so limited, the potential genetic resources are incalculable. Wild genes are mainly used to improve both plant and animal crop strains and as the basis for pharmaceuticals. Over half of all medicines at present prescribed have their origins in these genes. The world's food and the world's health have depended on the fecundity and variety of the wild.

Post-Lamarckian theories of the biosphere tend to veer towards the animistic (we have only to think of what Hampstead makes of Lovelock's Gaia Principle: nothing but russet-coloured clothes and earnestness), yet the recognition of ecological inter-connectedness must always lie behind any discussion of natural resources. This recognition is important because in considering genetic resources or biodiversity attention is drawn always to their richest sources – tropical moist forests, coral reefs and mangrove swamps. The importance of all three lies not simply in their range of species, genes and chemicals, nor, incidentally, in that they provide shelter for many unique groups of people, but also in their multiple ecological roles. They are carbon sinks and so help to prevent global warming; they affect currents and climate; forests in particular have a major bearing on the world's fresh-water supplies; they are sources of food and fuel not only for the people who live in or on them, but for much of the rest of the world too.

Half of even the most modest estimate of the number of species on earth are to be found in the tropical moist forests, which together occupy about 6 per cent of the world's land area. All but a fraction of that 6 per cent lies in the Third World. When we think of world genetic resources it is important to realise that those countries fortunate enough to have such forests and to be able to preserve them ought to be in a strong economic position. The value of, for example, the US pharmaceutical trade based on wild genetic and chemical resources is around US$14 billion a year; the world-wide figure is probably three times that amount.[6] Yet, once again, we find a story not unlike that told in the last chapter: the resources may well lie in the South but the control of research, technology, manufacture and markets is Northern. So it is that some states in the South, particularly in connection with Northern demands on developing countries to play their part in controlling global warming, have tried to use their possession of wild resources to wring concessions out of developed countries. One Northern response to these attempts has been to use biotechnology as a counter, so that the wild is no longer so important.

All our domestic animals and plants came, as we did ourselves, from the wild and the process is conscious, ancient and continual – we have only to think of the fairly recent domestication of the oil palm. It was the rediscovery, in 1900, of the work of the nineteenth-century scholar, the Abbé Gregor Mendel, which led to the most exciting and important biological discovery of this century, that the genetic information for inherited characteristics is contained in the substance deoxyribonucleic acid, known to most of us as DNA. This discovery forms the basis of the contemporary science of genetics and of what is now often called biotechnology.

Some simple figures may be used to illustrate the importance of genetic resources to the North. Between the years 1930 and 1975, in the USA alone, wheat yields per hectare increased by 115 per cent, those of rice by 117 per cent, maize by 320 per cent, sugar-cane by 141 per cent, peanuts by 295 per cent, soybeans by 112 per cent, cotton by 188 per cent and potatoes by 311 per cent. Roughly 50 per cent of these increases were a consequence of the use of genetic resources to improve strains and to render them more resistant to disease.[7] Similar startling results have been observed in milk production, where the use of genetics has also played a significant role. The resources themselves are not all wild; the concept of 'gene pools' is important and has recently come to mean all the genes available within inter-breeding groups of both domestic and wild plants and animals. In general, wild genes are used only if the characteristic required to improve the strain is sufficiently rare to warrant it and if the transfer is not too difficult, but they are crucial in the formation of primary genetic pools.[8] The preservation of the remaining diversity in the wild is a matter not only of continuing to improve existing domesticated species, but of preserving the options available to our children.

Not all agricultural improvements based on genetic resources have been confined to the North. The famous 'Green Revolution' of the 1960s and 1970s, which involved the widespread planting of high yield varieties (HYVs) of grains in the Third World, despite its devastating effects on poorer farmers (see Chapter 6), its methods and some of its results, did increase food production and was the product of genetic engineering. That increase had effects beyond the ruin of numbers of small farmers; because it demanded the dedication of large tracts of farmland which had been part of a traditional agricultural ecology, many genetic resources were destroyed in order to make way for the HYVs. The story, however, is one of management dominated by Northern, governmental or private rapacity and of the manipulation of supplies, processes and land in the interests of large landowners. It does not follow that genetic engineering in itself is bad, it is entirely a matter of whose interests are safeguarded. For the future, in the face of vastly increased populations and growing desertification, which could be exacerbated by climate change and the continued importance of agriculture in Southern economies, genetic improvements, as opposed to largely destructive chemical fixes, may come to be essential.

Genetic engineering may hold the answer to many of the world's problems, but it would be foolish in the extreme to suppose that the diversity of genetic pools can simply be maintained whether or not the naturally occurring genes survive. Quite apart from the attack on people's livelihoods as a consequence of the destruction of many of the sources of biodiversity, the world is unquestionably rapidly losing major resources. Clearer cases of meeting some very special sectoral demands by compromising the ability of future generations to meet their needs are hard to find. The destruction of the mangroves, of the coral reefs and of the tropical moist forests are all perfect examples of 'unsustainable' development.

Mangroves, of which some 14 million hectares throughout the world remain, are immensely complex and rich ecosystems. For example, some 2,000 different species of fish depend for their existence on them, as do innumerable invertebrates and plants.[9] Further, 90 per cent of the world's fish harvest, calculated by weight, breed in the mangroves.[10] Mangroves also play a crucial role in coastal preservation, even in land reclamation as their roots consolidate the mud in which they grow. Used sustainably, they are sources of timber, pulpwood, charcoal and fuel, but, as in so many other cases, they are not being used sustainably. Many of them are dying because river flows have been drastically reduced by damming for hydro-electric projects, others are being cut down for agricultural land, some are being destroyed to make paper, more are suffering from population pressures and from pollution caused by new industrialisation (see Table 3.1). In the case of Mozambique, and possibly Angola, the South African wars of intervention, in driving people to 'safer' areas, have also increased the pressure on mangroves.

Coral reefs are vanishing because, among other reasons, they are being mined for building materials, chemicals and even ornaments, because the seas are rising and because pyromaniacal Europeans are exploding nuclear bombs on them. The seas, unlike the forests, are 'commons', so there is a sporting chance that some fairly sensible agreements about them may emerge in the wake of UNCED and its successor conferences. Besides, the preservation of the greatest of all coral growths, the Great Barrier Reef, is very much in the interests of at least one developed nation, Australia. The destruction of that reef would not only mean the extinction of many marine species, both animal and vegetable, but also extensive damage to Australia's generally low-lying coastline. Such damage could lead, as well, to the destruction of what remains, after their pillage by Australian industry, of the great rainforests of northern Queensland.

It is possible that another and more insidious threat to coral reefs is emerging as a result of global warming. In recent years observers have noticed an increase in the phenomenon of coral 'bleaching' on reefs in the Caribbean and the eastern Pacific. Coral gets its characteristic colour from the presence of symbiotic algae, without which it will die. Pollution and sedimentation are both known threats to these algae, but in some places a connection has also been established between their loss and a rise in the surrounding water temperature of one or two degrees over what is usual. At the moment the rise in temperature is probably caused by environmental

Table 3.1
Loss of mangroves in selected countries since pre-agricultural times

Location	*Current sq km*	*percentage lost*
Africa		
Angola	1,100	50
Cameroon	4,860	40
Côte d'Ivoire	640	60
Djibouti	90	70
Equatorial Guinea	120	60
Gabon	1,150	50
The Gambia	510	70
Ghana	630	70
Guinea	1,200	60
Guinea-Bissau	3,150	70
Kenya	930	70
Liberia	360	70
Madagascar	1,302	40
Mozambique	2,760	60
Nigeria	12,200	50
Senegal	420	40
Sierra Leone	3,400	50
Somalia	540	70
South Africa	450	50
Tanzania	2,120	60
Zaire	1,250	50
Central America		
Guatemala	500	60
Asia		
Bangladesh	2,910	73
Cambodia	156	5
India	1,894	85
Indonesia	21,011	45
Malaysia	7,310	32
Pakistan	1,540	78
Philippines	777	61
Thailand	191	87
Viet Nam	1,468	62
Total	***76,939***	***56***

Source: World Resources Institute in collaboration with UNEP and UNDP, *World Resources 1992–93*, p. 178, Oxford University Press, New York and Oxford 1992.

stress, but the scientists who have made this discovery point out that if global mean temperatures do rise in conformity with current predictions, then the effect on coral reefs could be devastating.[11]

It is in the destruction of the forests that the threat to the greatest part of the world's genetic stock lies, and this is a question that we must also consider in the next chapter. This destruction (see Table 3.2) is a direct product of the mining approach to the wild and, yet again, the market is provided, at least partly, by the Northern appetite for forest products. Examples run from the Japanese exploitation of Chilean forests for special pulp for computer paper to English company directors' vulgar desire for mahogany boardroom tables. That sort of thing, together with grandiose and usually ill-designed and poorly constructed hydro-electric power stations and their attendant, swiftly silting-up reservoirs, the extraction by the cheapest means possible of substantial mineral deposits, the destruction of forests for mono-cultural plantation, all of which is financed by Northern banks in the name of growth, make a deadly combination. Poverty, debt and plans for growth inextricably linked to Northern trade are, as with everything else in cases of unequal development, at the heart of the destruction.

Before the UNCED conference three major agreements had been reached which were designed to protect the world's biodiversity. They are certainly better than nothing, but they neither go far enough nor address the problem of equity. One excuse for this last failure is to be seen in that curious survival, the concept of the 'tragedy of the commons' in which individual and social needs come into conflict and lead to the decline of the common good. It is, as Ben Wisner has observed,[12] a bad piece of social science and is really based on a failure to understand the radical differences in systems of land use. In societies based on commonage the administration of the land and its law is customary and generally has elaborate defences against destructive encroachments. The world of industrial capital does not legislate in this way at all and is uncomfortable with the idea of 'commons'. It is happier with property rights and only where these infringe some greater right are they modified. Many environmental and developmental agreements have property rights in mind and because of this they often run into trouble. Land can be grabbed and enclosed for exploitation or conservation – all this is harder when water and gas are at issue and the difficulty is reflected in such conventions as exist on the seas and the atmosphere.

Table 3.2

Preliminary estimates of tropical forest area and rate of deforestation for 87 tropical countries, 1981–90 *(million hectares)*

Regions & sub-regions	Number of countries studied	Total land area	Forest area 1980	Forest area 1990	Area deforested annually 1981–1990	Annual rate of change (per cent)
Total	*87*	*4,815.7*	*1,884.1*	*1,714.8*	*16.9*	*-0.9*
Latin America	*32*	*1,675.7*	*923.0*	*839.9*	*8.3*	*-0.9*
Central America & Mexico	7	245.3	77.0	63.5	1.4	-1.8
Caribbean sub-region	18	69.5	48.8	47.1	0.2	-0.4
Tropical South America	7	1,360.8	797.1	729.3	6.8	-0.8
Asia	*15*	*896.6*	*310.8*	*274.9*	*3.6*	*-1.2*
South Asia	6	445.6	70.6	66.2	0.4	-0.6
Continental SE Asia	5	192.9	83.2	69.7	1.3	-1.6
Insular SE Asia	4	258.1	157.0	138.9	1.8	-1.2
Africa	*40*	*2,243.4*	*650.3*	*600.1*	*5.0*	*-0.8*
West Sahelian Africa	8	528.0	41.9	38.0	0.4	-0.9
East Sahelian Africa	6	489.6	92.3	85.3	0.7	-0.8
West Africa	8	203.2	55.2	43.4	1.2	-2.1
Central Africa	7	406.4	230.1	215.4	1.5	-0.6
Tropical Southern Africa	10	557.9	217.7	206.3	1.1	-0.5
Insular Africa	1	58.2	13.2	11.7	0.2	-1.2

Source: FAO of the United Nations, 1991, quoted in World Resources Institute, *World Resources 1992–93*, p. 119,Oxford University Press, New York and Oxford 1992.

The world's first conservation treaty, signed in Ramsar, Iran in 1971 is usually called the 'Ramsar Convention', though officially its somewhat less snappy title is the Convention on Wetlands of International Importance especially as Wildfowl Habitat. It came into force in 1975; by 1990 fifty-eight nations had signed it and, between them, had nominated over 500 wetland sites, comprising more than 30 million hectares, as protected areas. The Convention is overseen by the WWF, the IUCN and the International Wildfowl Research Bureau. Admirable as it is, it is nonetheless very restricted in its aims.

Another, very controversial, agreement is CITES (Convention on International Trade in Endangered Species of Wild Fauna and Flora). Agreed in 1973, it, too, came into operation in 1975 and is managed through a system of import and export licences. It has been more successful in the case of animals than of plants in reducing the destructive trade in endangered species between its member countries, of whom there are 103. It can do nothing to prevent the trade continuing among non-members. Among its provisions is an agreement that trade may be forbidden not only in the animals and plants themselves, but also in products derived from them. Controversy over its implementation was widely reported when its provisions were used in an attempt to halt the decline of the African elephant by banning the ivory trade. World concern had been aroused by the decline of the elephant population in Africa from 1.2 million to around 600,000 in about eight years. Poaching in Kenya was the focus of media coverage, but it was argued, particularly by Zimbabwe and Zambia, that a total ban was unnecessary in the case of those countries which could show either stability or a positive increase in the numbers of elephants (the increase in elephant numbers in those two countries and in the forests of Gabon and Congo may simply have been due to improved counting methods).[13]

The Convention on the Conservation of Migratory Species of Wild Animals (the Bonn Convention) has been signed by thirty-four states and is designed to protect any species of which some or all members migrate, crossing international boundaries, even if they are merely transient. For this convention to have a substantial effect, many more countries must agree to it. There are a number of regional agreements as well, of which the Berne Convention on the conservation of European wildlife is one example. It is also worth mentioning a protocol formulated by UNEP. This is called the International Undertaking on Plant Genetic Resources and is directed towards their conservation.[14] UNEP is lobbying its member countries in an attempt to have this protocol widely adopted.

Excellent in their way as these agreements are, what was called for from UNCED was a new and major convention on biodiversity aimed at protecting all the important areas and their enormously varied species. But any such convention will fail to achieve its ends unless the parties to it confront the issue of ownership and it was this confrontation which was so strenuously resisted by the US and, to a lesser degree, some other interested industrialised nations. Foremost is the question of the right of the forest-dwellers and all those who, in sustainable ways, depend on the forests for their livelihoods, to conduct their own affairs without disruption and free of attack. This must be allied to the right of these people fully to participate in any exploitation of the forests and their resources. Unless these rights are established and efficiently safeguarded any agreement will be worthless. It is over this issue that President Clinton, despite his improvement on the Bush record, is being most cautious.

The point was not entirely lost on the framers of the Convention on Biodiversity. In the preamble, for example, we find two references which suggest that hearts may sometimes be found in the right place. They assert, among many other things, that the Contracting Parties

> *Reaffirming* that States have sovereign rights over their own biological resources ... *Recognizing* the close and traditional dependence of many indigenous and local communities embodying traditional lifestyles on biological resources, and the desirability of sharing equitably benefits arising from the use of traditional knowledge, innovations and practices relevant to the conservation of biological diversity and the sustainable use of its components,

agree in the document to a whole host of wonderful things, but, as we shall see, few of them offer much of substance.

If we turn to *Agenda 21*, we find a similar story. Paragraph 2 of Chapter 15, offers an almost idyllic account:

> Our planet's essential goods and services depend on the variety and variability of genes, species, populations and ecosystems. Biological resources feed and clothe us and provide housing, medicines and spiritual nourishment. The natural ecosystems of forests, savannas, pastures and rangelands, deserts, tundras, rivers, lakes and seas contain most of the Earth's biodiversity. Farmers' fields and gardens

> are also of great importance as repositories, while gene banks, botanical gardens, zoos and other germplasm repositories make a small but significant contribution. The current decline in biodiversity is largely the result of human activity and represents a serious threat to human development. (paragraph 2)

Scarcely, one would have thought, the stuff to cause conniptions in the president of the USA and there is only one other paragraph which might strike a nervous White House as a little close to the bone. The document suggests that all states should

> Recognize and foster the traditional methods and the knowledge of indigenous people and their communities, emphasizing the particular role of women, relevant to the conservation of biological diversity and the sustainable use of biological resources, and ensure the opportunity for the participation of those groups in the economic and commercial benefits derived from the use of such traditional methods and knowledge. (paragraph 4, subsection [g])

In the Convention itself this emerges as:

> Subject to its national legislation [each state shall], respect, preserve and maintain knowledge, innovations and practices of indigenous and local communities embodying traditional lifestyles relevant for the conservation and sustainable use of biological diversity and promote their wider application with the approval and involvement of the holders of such knowledge, innovations and practices and encourage the equitable sharing of the benefits arising from the utilization of such knowledge, innovations and practices. (article 8, clause j)

This, and similar pieties, refer repeatedly to the ownership of genetic resources by the states in which they occur and demand respect for the undefined 'rights' of indigenous people. They also refer to the importance of the traditional knowledge of the *indigènes*, but all this has to do with conservation. The development and use of genetic and other material thus conserved is the stuff of biotechnology. Here the treaty refers extensively to the need to transfer technology, but hedges it with the obvious, if fatal, condition:

> Access to and transfer of technology referred to in paragraph 1 above to developing countries shall be provided and/or facilitated

> under fair and most favourable terms, including on concessional and preferential terms ... In the case of technology subject to patents and other intellectual property rights, such access and transfer shall be provided on terms *which recognize and are consistent with the adequate and effective protection of intellectual property rights* . (Article 16, clause 2; emphasis added)

The Convention does go on to suggest that these property rights should not be allowed to run counter to the purposes of the treaty. But it fails to suggest serious measures designed to ensure that the poor countries that 'own' the genetic sources or indigenous people whose knowledge is so important in the development of many of those privatised resources do not, as has always been the case so far, have to pay through the nose for that biotechnology. It must also be made possible for the Third World countries committed by agreement to the preservation of species and wild genetic resources to have marketing control over them. We shall deal with the issue of existing marketing restraints on developing countries in a subsequent chapter, but the present world pattern does not offer much hope either for the preferential transfer of technology and 'intellectual property rights' or for marketing arrangements which will allow the poorer nations to develop and process their own resources.

Inevitably such an agreement will mean giving, yet again, serious attention to the issue of indebtedness. Recently, the Northern understanding of globalism, opportunist to the last, has introduced 'debt-for-nature swaps'. Year by year loans attract interest in line with the originally agreed rates, but the lenders, who treat the loans as stock-in-trade, write them down in value every year. This device preserves them from financial disaster if the loans are never repaid. A number of interested Northern bodies have been buying these loans from the lenders at the written-down value and allowing the borrowing governments to repay the debts, at an agreed face value, in unconvertible local currencies. These funds are then used in the country concerned in nature conservancy projects. In the largest of all these agreements so far, a consortium of the WWF, the Nature Conservancy and the Missouri Botanical Gardens bought from the banks US$10 million of Ecuador's debts relatively cheaply. The funds thus realised from the Ecuadorean government are being used 'To support mangement plans for protected areas; to develop park infrastructure; to identify and acquire small nature reserves; to fund species inventories through a Conservation Data Centre. Priority target areas are six Andean and Amazonian national Parks'.[15] There have been a number of other such deals in Latin America, Africa and South-East Asia and one in Europe (Poland). In each case the governments of the developing countries are provided with expert advice on policy and practice, often by the groups concerned. We should not, I suppose, be startled to discover that in none of these swaps are the rights of those ordinary people who depend on the forests ever mentioned. An even more blatantly colonising deal is to be seen in the 'debt-for-equity swaps' in which private interests exchange the cheaply purchased debts for shares in Third World businesses. Mahatma Gandhi went on hunger strike about this kind of thing. However, the impact of these arrangements on the general level of indebtedness and, indeed, on conservation, is relatively tiny; the face value of debt so far thus redeemed is US$98,881,860 at a total cost of $17,271,141, which has generated $61,491,133 for conservation projects.[16]

All agreements are the product of politicking, bargaining and compromise and many developing countries ought to be in a position to force concessions from Northern governments by virtue of their possession of species and genetic resources. The growth, particularly since the 1980s, of biotechnology has, however, seriously weakened the bargaining power of the South. Scientists have learnt how to modify the DNA of organisms without having to depend on external genetic characteristics to do so, although they still tend to regard the existence of that diverse natural genetic resource as a sort of back-up to their activities. Effectively they are in the business of producing 'nature' and many of them tend to feel that there is little point in troubling about the wild so long as clean and neatly packaged production processes are possible in the laboratory.

Clearly there are huge benefits to be derived from advances in biotechnology. It offers, though in some instances in the very long term, the possibility of environmentally harmless ways of dealing with crop pests; with disease, both animal and human; it may provide safe, efficient and easily controllable methods of family planning; it may well come up with solutions to the problems of hazardous waste. It is, indeed, the late twentieth to early twenty-first century super-science, it kills all known germs – Jules Verne should have seen this. Properly applied and using the maker's directions it could undoubtedly improve matters substantially for developing countries, but the makers are in the developed world. Research in this, as in every other, field has to be paid for. Universities, state and industrial research institutions are all very expensive to run and the customary way of recovering costs is to launch successes on to a market. One recent and bizarre illustration of this was seen in the successful attempt by a laboratory in the US to patent a mouse bred to be more susceptible to cancer and known as 'onco-mouse'.[17]

From our point of view it is programmes of this kind which point up the weakness of the developing world's position. A monstrously expensive and somewhat isolated research industry, financed partly by state, but commonly by market-oriented corporations, depends for its existence on its place in the markets and is busy engineering a whole new range of products which could, lunacies apart, enhance world development. Although this industry is still, to a degree, dependent on wild genetic resources, that dependence is increasingly remote. A technology which could substantially improve the lot of the poor will be incorporated into the free-market system which is already so heavily weighted against the Third World. In this case the immense species and genetic resources which could have been bargaining points are, for the time being at least, now seen by the North as of less importance. One makeweight is sometimes limply inserted into the argument – it is said that aesthetic and spiritual values should encourage the preservation of the wild. This may well be, but it is a suggestion that seems not to carry far on the golf course at Camp David. But, above all, the industry which is patenting new genetic material has not only directly reduced the value of a Southern resource, it has only been able to do so because it raided that resource in the first place. In these patents we may thus see yet another transfer of resources from the poor to the rich.

A separate and immensely important consideration is that of safety. Uncertainty about the completely unpredictable effects of genetically engineered species on existing domesticated and wild organisms is one major fear, but there is an even greater issue. Northern corporations are not well known for their concern for the well-being of their Southern customers. Levels of industrial safety may best be illustrated by the explosion at the Union Carbide factory in Bhopal, India, in 1984 which resulted in the deaths of thousands of people. Chemicals requiring very careful handling and proper protection for their users are frequently made available in the Third World with inadequate instructions and no regard to the improbability that workers will have adequate protective clothing. One of the most alarming examples of apparent corporate unconcern for human safety was to be seen in the continued pushing on Third World women of formula milk in preference to breast-feeding long after it was widely recognised that the latter was substantially better and safer for babies. None of this encourages the view that Northern companies marketing biotechnology in the developing world can be relied upon to do so safely.

Before the great meeting, UNCED's third preparatory committee had asserted the links between the need to conserve species diversity and a host of other issues: climate change, forests, oceans and coastal areas, land resources, fresh water, population, environmental education and biotechnology. It proposed the establishment, by 1995, of a major programme to assess the extent of the world's biological resources, to maximise and to spread their benefits, to improve conservation and to promote education, training and institution-building for the management of these resources. Targets were suggested for this programme: by the year 1995 to have made biodiversity an issue in the 'mainstream of national and international policy-making'; by 2000 to get on the road national surveys, co-ordinated with a global survey, designed to establish 'baseline assessments of biological resources'; to 'stabilize the depletion of biological resources at natural levels in non-tropical regions by the year 2010 and in tropical regions by the year 2025'; throughout all this the programme should also be promoting 'the use of biological resources to enhance sustainable development'.[18]

Some may feel that, at the present rates of destruction, the task of assessing biological resources may be considerably easier by the year 2025 than at present. This fear becomes more pressing when we look at what was actually accepted by the participating states in *Agenda 21* and the Convention on Biodiversity, in both of which most of the objectives have been preserved but all mention of target dates has been carefully expunged. Lacking these, apart from repeatedly emphasising the importance of conserving diverse species both, as they put it, *in situ* and *ex situ* (all those botanical gardens and farmers' fields mentioned in the *Agenda*'s Chapter 15), exchanging information, transferring technology (so long as due regard is paid to intellectual property rights) and institution-building, what UNCED quite correctly stressed is the need for education. Northern and Southern politicians, bureaucrats and publics must understand the realities of biodiversity and the need to preserve as much as possible of what remains.

It would be foolish simply to decry UNCED's efforts, even though they may be oddly directed. In itself the Convention is so weak because it is designed to straddle horses running in opposite directions. On the one hand it strives to reduce the rate of species destruction by demonstrating that more money can be made not by destruction but by intelligent management; on the other, its nod in the direction of the rights of women, children, indigenous farmers and the states in which they live runs counter to the naturally exploitative instincts of the market. However, there are few international conventions which do not have muddled aims and their value can lie in their stage-setting functions. Once *in situ* they may be used to go further.

Even weaker is UNCED's cautious approach to biotechnology, which it defines as

> the integration of the new techniques emerging from modern biotechnology with the well-established approaches of traditional biotechnology. Biotechnology ... is a set of enabling techniques for bringing about specific man-made changes in deoxyribonucleic acid (DNA), or genetic material, in plants, animals and microbial systems, leading to useful products and technologies. (Article 21, Chapter 16, paragraph 1)

The aims of *Agenda 21* were more or less subsumed into the Convention and are clear enough:

> By itself, biotechnology cannot resolve all the fundamental problems of environment and development, so expectations need to be tempered by realism. Nevertheless, it promises to make a significant contribution in enabling the development of, for example, better health care, enhanced food security through sustainable agricultural practices, improved supplies of potable water, more efficient industrial development processes for transforming raw materials, support for sustainable methods of afforestation and reforestation, and detoxification of hazardous wastes. (chapter 16, paragraph 1)

Once again, in meetings before the actual conference, targets were set: plant and animal productivity to be increased by 25 per cent by the year 2000; a 25 per cent reduction in the use of pesticides by the same year; an unspecified increase in the productivity of marginal land by the use of nitrogen fixation and mycorrhiza and a reduction in the use on them of chemical fertilisers. All these dates had vanished in the final documents.

So far as 'better health care' is concerned UNCED suggested that health for all by the end of the century should be achievable, though not simply by health programmes. The provision of safe water and food, of good sanitation systems, would be as important as the widespread availability of essential drugs and programmes of universal immunisation. What the Rio documents look forward to in a misty future, among other things, are biotechnological contributions to diagnostics, the development of therapeutic and growth-promoting agents, new forms of birth control and the development of ways of detecting and curing genetically transmitted disease and malformations. In view of Northern experiments in 'social hygiene' earlier in this century, we might, perhaps, view this last suggestion with some reserve. Even so, we must applaud UNCED's optimism on behalf of our new wonder-science.

Optimism, perhaps not completely misplaced, crops up in another guise in the options for the Earth Summit:

> Biotechnology also offers new opportunities for global partnerships, especially between the countries rich in biological resources (which include genetic resources) but lacking the expertise and investments needed to apply such resources through biotechnology and the countries that have developed the technological expertise to transform biological resources so that they serve the needs of sustainable development. (*Agenda 21*, Chapter 16, paragraph 1)

Once more, an admirable sentiment, and this time one which we might, with some confidence, expect to see embodied quite soon in relationships between the corporations of the North and those nations able to preserve some of their genetic riches. What the texts cannot do, of course, is to determine the nature of these partnerships. It may well not have been beyond the competence of UN officials to work out rules ensuring that the developing countries do not simply wind up as junior partners completely in the control of their Northern 'seniors' and of Northern markets, but to have done so would have ensured the non-attendance of the USA.

The Earth Summit sees the preservation of biodiversity primarily as an environmental issue so that questions to do with the ownership and control of resources are repeatedly lost in a haze of generalities. So far as UNCED is concerned, biotechnology offers bright new technological answers to some tiresome problems and apart from a few worries about its safety and environmental soundness it is merely an adjunct to environmental management. What should be at stake here is development. In a world in which biological resources are managed by and for the people who live among them environmental destruction is substantially less likely. In a world, also, in which research and information are not automatically privatised and put up for sale then biotechnology might also be reasonably well and safely managed. The weakness of the biodiversity/biotechnology Convention flows directly from UNCED's lack of developmental priorities.

4

Forests and Forestry

There is a distant parallel to be discerned in the shift of gravity in the UNCED process from development to environment and the shift in emphasis from forests to biodiversity. Both, consciously or otherwise, contrive to move the matters further into concern for things and further away from the immediate needs of people. That aside, it proved possible, in spite of a number of contentious areas to achieve a Convention, no matter how weak, on biodiversity but not on forests for which UNCED kept to a general statement of principle. Its unlovely title is 'Non-legally Binding Authoritative Statement of Principles for a Global Consensus on the Management, Conservation and Sustainable Development of All Types of Forests'. We shall call it the 'Statement of Principles'. Some reasons for producing nothing stronger than this general and non-committal document are, perhaps, to be found in the first paragraph of the preamble to the Statement: 'The subject of forests is related to the entire range of environmental and development issues and opportunities, including the right to socio-economic development on a sustainable basis.' If the Convention on Biodiversity seemed threatening to US interests, then any agreement beginning with that position would have been doubly so and, we suspect, the USA would not have been the only Northern state to have rejected it. So it is that biodiversity and its forestry connections are the subject of a Convention, but forests, and consequently those who depend on them, are not.

A careful reading of the Statement leads one to suppose that its authors faced a considerable problem of definition. They talk, reasonably, of the application of the principles to all kinds of forest and to those people and economies affected by and adjacent to them. They do not say that they mean, in addition, all other social forms of arboriculture, but we may be generous and assume that this was their intention. The Statement calls for accurate and open information, discussion and consultation with local and indigenous people, together with recognition of their rights; the full acceptance and understanding of the role of women in many of the most crucial areas of forestry; the 'promotion of sustainable patterns of production and consumption' (paragraph 7[a]); understanding and accepting the 'relationship, where it exists, between the conservation, management and sustainable development of forests' (paragraph 6 [b]). It calls for finance for all this, for scientific research, for reforestation and for developing countries to be able to take sustainable control of their own forest resources. It goes on to suggest non-discriminatory trading arrangements, the provision of better market access and the arrangement of better prices in favour of the developing countries. We need not be astonished that the North, together with some of its allied Southern élites, applauded the idea, but neglected to provide an agreement.

The need to conserve biodiversity led the contracting parties to concern themselves with the conservation of tropical moist forests, though to be sure the *indigènes* and, let us not forget, women must be considered too. In their Convention the Northern states who agreed to it thus met some of the anxieties of their constituents, anxieties which are catered for in those familiar advertisements: 'Feel good. Buy yourself a square metre of tropical rain-forest and save the world.' There is a pre-packaged messianism in many such offers (some for forests, some for children – and then there is the 'adopt-a-granny' offer) which successfully diverts attention not only from the real political and economic issues but even from, in the case of the appeals on behalf of rainforests, the importance and place of forests and forestry in general. We are inclined to the jaundiced view that the tropical forests were seen by the panjandrums of UNCED as an easy area of agreement because only a small number of people live in them and, for most of us, they are far away.

Nevertheless, anxiety about the fate of tropical forests is not misplaced (see Table 3.2), their importance in the world's ecology is well known even if the livelihoods of those who, in sustainable fashion, have depended on them for centuries are not too clearly understood. We have already sketchily set out, in the last chapter, the position in which developing countries find themselves. Northern actors in the matter, recognising that their long-term interests may not be best served simply by *laissez-faire* approaches of the grab-as-grab-can kind, set up in Geneva, in 1983, an agreement known as the 'International Tropical Timber Agreement'. This led to the formation in 1985 of the International Tropical Timber Organization (ITTO), which has its offices in Japan and is made up of representatives from producing and consuming countries and of the timber trade. Both the agreement and the ITTO have accepted the principle that tropical forests must be used sustainably and conserved and this acceptance should not lightly be dismissed as window-dressing, even though one might raise an eyebrow at the wisdom of setting foxes to guard chicken-coops.

There are a number of objectives in the ITTO programme which, however, could lead the unwary to suppose that it is little more than a trade organisation. Research and development in forest management and in the uses of wood, reforestation, the establishment of modern timber- processing facilities in producing countries, and the provision of information designed to increase trade in timber are among the most important of these. Sustainability, in this context, has more to do with keeping the barrel rolling than with some other central issues. The concerns shared by the members of the ITTO do not readily run to the rights of indigenous forest people and others who depend on forests for their livelihoods, though they acknowledge that these are important; nor do they run to the problems of biodiversity except in recognising that, in the words of one of their most distinguished advisers, 'Special measures need to be taken to preserve the intra-specific variation of species of economic importance'.[1] Both admissions are engagingly mimetic of ITTO members' interests.

Another group has also entered the arena, this time composed of the World Bank, UNEP, the Food and Agriculture Organization of the UN (FAO) and the World Resources Institute (WRI). In 1985 it launched the Tropical Forestry Action Plan (TFAP), the aims of which would seem to be wholly beneficial.[2] By calling for a thorough review of the state of tropical forests, country by country, in terms of the needs of the people and of the environment and in terms of their economic value, it hoped to provide the basis for developing countries to set their own priorities. The TFAP is also committed to the sustainable exploitation of forest resources with the intention of making the lives of the poor more tolerable. It could be seen as the 'aid and development' programme to be set alongside the 'trade and development' purposes of the ITTO. That it has not succeeded in its aims and that its national action plans are frequently dominated by timber-trade interests, so that it is often seen as a 'loggers' charter', is probably due to its somewhat compromised origins. Its founding group could not be expected to agree that pressures on forests produced by landlessness or poverty owe much of their origin to the political and economic concerns of Northern financial and state institutions and of their allied Southern oligarchies. A revised version of the Plan, to be called the Tropical Forest Action Programme, is to be launched now that the UNCED summit meeting is over. An attempt will be made to reform its content and administration so that it conforms a little more to the original intent of the Plan.

Nonetheless, behind these worthy attempts at self-regulation some Northern timber entrepreneurs have a rather more full-blooded approach to the weak and needy. A startling case in point was reported in a newspaper story in 1991.[3] It described the acquisition by the United Dutch Company, in which the now financially troubled Lord Beaverbrook had a majority share, of Guyana's state timber corporation, Demerara Woods. It was purchased together with the right to 'extract' timber from more than 2 per cent of that country's land area.The figures are instructive. Between 1982 and 1988 the EC, the World Bank and others lent Guyana some US$40 million, of which about three-quarters went to support Demerara Woods. In 1988 the hitherto Marxist government decided on a shift to a form of monetarism and enlisted Mrs Thatcher's help. The British made available, as aid to Guyana for 'economic reconstruction including privatization', further sums amounting to about $37 million in the years 1989–90 (previous annual totals had reached their maximum in 1988 at $1.1 million).

A number of British companies sniffed around Guyana's national assets, but the greyhound among them proved to be Lord Beaverbrook, a man who should have gone far. Together with a consortium of other business people he snapped up Demerara Woods for the unrepeatable 'wrap it up and take it home' price of $16.5 million, somewhat less than the amount that had been invested in it. The price included a fifty-year timber concession on 440,000 hectares of Guyana's forests. Demerara Woods, renamed Demerara Timber, was then sold by Beaverbrook to the company in which he held a majority interest, the United Dutch Group, for about $104 million. The Group had valued the company at $126 million, while also accepting a modest valuation for half of the concession alone of about $135 million.

If UNCED and its heirs and assigns are to arrive at a treaty on forests they have not only to deal with the problems raised by the Northern enthusiasm for turning an honest penny, but also with those caused by contemporary development practice and received wisdom. Development funds are commonly administered through development banks and they operate on roughly the same principles as ordinary commercial banks. An official in one of them remarked recently to one of the present authors that his bank could not make commercial sense of forest management unless it involved replacing felled indigenous species with fast-growing species, principally eucalyptus. In the financial arrangements between the developed and the developing worlds, which are usually dictated by the richest partners, commerce is all and its terms and probity are those of the balance sheet. Many of the demands made even by environmentalists, let alone demands from those urging solutions to the problems of unequal development, are, in the ordinary sense 'non-commercial'. We must grapple with the blinkered and essentially political view that 'commercial' and 'economic' are nearly synonymous.

The history of the destruction of boreal forests goes back at least 5,000 years, though it accelerated during European maritime expansions, the Industrial Revolution and those first steps in agribusiness, the land enclosures. Add to this history the Northern style in forest financial management and Southern scepticism about Northern plans for forest preservation can scarcely astonish. Clearing boreal forests usually exposed rich, or at least enrichable, agricultural land; clearing tropical forests usually does not. However, this argument cannot weigh against the price that Northern industry is prepared to pay the Third World simply for 'mining' its forests. That price may well be a long way below its proper value, but it is better than nothing and pressure to service debt, often from those very Northern financial institutions advocating preservation, will persuade countries with low incomes to accept it.

Brazil's case is different. It is the state in which the largest part of the world's tropical moist forest is to be found, and it leads the way in resistance to any agreements to do with forest preservation and which might interfere with its right to do as it wishes with what it sees, with some justice, as its own resource. We have already pointed to the reasons why the governments of developing countries might not welcome Northern initiatives, but Latin America is special. More than any other part of the colonised world, it was treated by its invaders as a treasure trove for their own enrichment and the enrichment of their European masters. All Latin American economies were built on the principle of extraction, mainly to finance the elaborate costs of European states, and this formed a habit which has survived in most of Central and South America ever since. Brazil is rapidly becoming part of the industrialised world; its ruling oligarchy and its provincial satraps together with an increasingly powerful middle class are recreating, for themselves and in their own fashion, the mores of the richer parts of the USA. At the same time the Brazilian population of the under-paid, poor and absolutely destitute remains vast; 'trickle-down development' theories, always fairly suspect, are plainly not working. It is futile to expect Brazil's rulers willingly to accept agreements which they will inevitably see as compromising their interests for anything less than some other equal gain.

ITTO estimates for 1985 suggest that about 828 million hectares of what it describes as 'productive tropical forest' remained, about 40 per cent of them in Amazonia.[4] Duncan Poore, in his report to the ITTO Council in 1989, pointed out that at the very most the area of these forests under sustained yield management, throughout the world, amounted to 1 million hectares. We need not quibble with the methods and purposes of this kind of management, its scale is indicative of the lack of any serious policies. Timber extraction is, however, only one problem. The story of the destruction of tropical forest for the creation of range and farmlands of dubious quality is well known. Equally important is the degree to which the forests harbour industrial resources. Apart from sites for hydro-electric projects, largely unquantified deposits of fossil fuels and minerals are also tempting the clearers. This is particularly true of Amazonia, where large deposits of natural gas, oil, iron ore, bauxite, gold, nickel, copper and tin have been discovered.

We have given the current estimates for the rates at which tropical forest is being destroyed in Chapter 3. In very broad terms, the other major kinds of forest are open tree, tropical and temperate, and boreal. None of these is as rich in species as tropical, closed-canopy, forest and the last two differ in another, more important, way. Boreal and temperate forest ecosystems are far less sensitive to damage and far more resilient than tropical forests. Their soils are richer and extensive clearing for agricultural purposes has largely been successful. Nonetheless, in both northern and temperate zones forests and woods fulfil purposes beyond the harbouring of forests species, although those, too, are important. Among other things they affect climate, prevent erosion, assist in soil-formation and fertility and affect water-tables; they are extensively 'farmed' for their products; they are, of course, also carbon sinks, though the importance of that role may be exaggerated. Global figures for the management and protection of all natural closed forest are offered by the World Resources Institute (see Table 4.1), and even here we can see that a substantial amount of 'managed' forest lay in what was the USSR, where 'management', for the large part, probably meant little more than theoretical state ownership. Less than 50 per cent of the original European forest cover survives, the rest has been cut down to make way for agriculture or construction or fuel, though some planting, particularly in Nordic areas, is redressing the balance. The USA has destroyed over 95 per cent of its virgin forests within the last two centuries, a destruction accompanied by one of the world's more successful genocides. Given this dubious history of Northern 'mismanagement', appeals to the South to leave its forests standing ring a little hollow.

Table 4.1

Global forest management and protection

	Total closed forest (1980) (000 ha)	*Managed closed forest (1980) (000 ha)*	*Protected closed forest (1980) (000 ha)*
Africa	222,278	2,327	118,019
The Americas	1,212,849	102,884	53,573
Asia	424,713	48,705	25,050
Europe	136,652	74,628	1,752
USSR	739,900	739,900	20,000
Oceania	86,168	0	8,009

Source: extracted from World Resources Institute, *World Resources 1992–93*, Table 19.1, pp. 286–7, Oxford University Press, New York and Oxford 1992.

Estimates suggest that tropical open-tree formations are being cleared at the rate of about 4 million hectares a year. The most recent figures available were provided by the FAO/UNEP *Tropical Forest Resources Assessment* published in 1980 and covering the years 1976–80. Figures in this assessment, based on less closely observed surveys than those for the closed moist forest systems, show that nearly half the destruction takes place in tropical Africa, particularly in the eastern Sahel and in tropical Southern Africa. A further 1.3 million hectares is destroyed in Latin America and the remainder in Asia. Such forests in all three regions are suffering from over-exploitation for fuelwood, charcoal-making and grazing. In Africa and Asia they are also being cleared to make way for the growth of cash-crops like coffee, cocoa, palm oil and rubber.

Forests, woods and small stands of trees of all kinds have been a secure source of food and fuel and have provided valuable grazing for domestic stock since before recorded history, particularly in the South. With some relatively unimportant exceptions the relationship, until very recently, has been sustainable. It was largely and, to an increasingly beleaguered extent, still is a use of common resources. Once more the dreary tale of what has threatened this use of forest, of the people who live in and around it, is well known. Colonialists took little note of the ecology of the people they colonised, surrounded their conquests with artificial borders which frequently threatened forest and grazing patterns, seized the land, made of it private property, turned much of it into cheap sources of produce for Northern markets, introduced destructive mono-cultures and, not the least, first enslaved and then reduced to destitution vast numbers of people whom they claimed to 'protect' and 'civilise'.

There are no 'might have beens' in history. It is now impossible to disentangle the reasons for urbanisation, industrialisation, population growth and a host of other problematic questions from the legacy of colonialism – it changed the face of the world. What is certain is that despite the massive movement of people into the cities of the Third World a vast and frequently impoverished rural population will remain and will also increase. Even now it is a population struggling to keep the old and sustainable ways alive – not out of some mindless conservatism, but because they work. Many of these sustainable ways have to do with trees, whether or not in forests. UNCED seems to have introduced a false sectoral division by dealing only with 'forests' and then almost only as a part of land management. It is almost a cliché to suggest that this arbitrary reduction of the importance of trees to only one form of growth effectively reproduces colonial perceptions and leaves the way wide open for the interests of Northern commercial forestry to come first, in a sustainable fashion of course, within these notionally hermetic 'properties'.

Britain established forestry departments in Africa as early as the first decade of the twentieth century. Their purpose was to oversee and ensure the efficiency of the commercial exploitation of the forests in their command. For the colonisers forests were principally a source of wood, either for fuel, as in the case of the use by the British of Sudanese forest for the steamers working the Nile, or, as in the Gold Coast (now Ghana), for a variety of industrial purposes. It was essential to log as swiftly as possible to meet a rapidly growing demand and in a world where Europeans were 'the Lords of Humankind' little, if any, account was taken of the local people and their relationship with the forests. Indeed they were largely seen as illegal predators on a resource which for everyone's good, including their own, should be 'managed' and employed productively for commerce.

Foresters became the guardians of a 'national' resource, imposing fines on unlicensed cutting and raising additional revenue with cutting permits. The British were not alone in this sort of practice; in Mali (known in the colonial period as the French Soudan) the French expropriated fallow woodland for state forests, largely ignoring any customary land rights. They imposed onerous and arbitrary fines and created in the people a powerful hatred not only of the state, but of foresters and forestry reserves.[5]

Patterns of expropriation and the imposition of commercial 'order' in making a resource out of what had previously been living and working territory were introduced by all the colonising powers. Led by the experience of the British Industrial Revolution, colonisers brought an industrial model to their administration of the varied resources they sought to extract, timber no less than diamonds. Imperialism was overwhelmingly successful and, so far as the modern world is concerned, was inseparable from industrialisation. Its success rendered difficult, if not impossible, any view of progress which did not assume industrial forms, so it was not surprising that the new leaders of the nascent post-colonial states adopted the bureaucratic patterns of the past. As in every other colonial industry, governmental infrastructures for the management of forests had been created, which were taken over by the new states during the periods of decolonisation, and so a confluence of attitudes, if not always of interests, was achieved.

Prevailing attitudes also informed the thinking of the aid agencies. Thus it was possible for the significantly named Division of Forestry Products (now the Forestry Department) of the FAO to say of forests: '... essentially a wood producing unit ... its treatment must be conditioned by the technological properties of its products for their industrial utilisation'. This attitude was not totally unenlightened and the role of forests in other directions was recognised; in many places forests were planted for the purpose of preserving water supplies and catchment areas. Even these concessions to the wider role of forests did not detract from the central concern of forestry departments of all kinds, which concentrated on the production of wood for export and for the emerging wood-based industries. Foresters, whether dealing with natural forest or with their new plantations, were concerned solely with the industrial exploitation and industrial 'farming' of the forest. Could this be the first known case of people unable to see the trees for the wood?

Prompted by aid workers who began to agitate about the question, the North stumbled across the 'woodfuel crisis'. Domestic fuel is commonly wood and it became, for a variety of reasons, increasingly difficult to obtain. Women were being compelled to spend ludicrous amounts of their already overburdened time in searching for it – in some cases between four and five hours each day.[6] For the North and its agencies trees imply foresters; the answer to the woodfuel problem was to plant fuelwood as quickly and efficiently as possible. This meant setting up plantations of trees which would burn well and grow quickly and could be cropped for the stove; preferably, for the sake of efficiency, planted in straight lines and decently fenced.

This strategy, simply because it failed to take account of the ways in which trees fitted into people's lives, met with total, sometimes even comic, opposition. Binar Agarwal[7] reports the case of Ethiopian labourers who planted saplings upside down, and there are many other instances of either direct resistance to, or simple neglect of, these ill-considered and industrially inspired fuelwood plantations. For rural people in most of the South, trees and hedges, whether 'natural' or planted, are an integral part of their farming management structures. They serve as fodder, for fruit, as protection against erosion; their place in relation to water supplies is recognised; they provide building supplies, shade and fuel. This last is not, as a rule, obtained by felling trees, but in the normal processes of decay and of pruning. The growing shortage of woodfuel was part and parcel of the increasing collapse of traditional farming which, as we have observed elsewhere in this book, was a consequence of a complex of post-colonial pressures.

To put this into another context we should remember that over 2,000 million people use wood for domestic heating and cooking: in some countries like Myanmar, Ethiopia and even oil-producing Nigeria, wood provides 75 per cent of all energy supplies. Deforestation on the scale at present horrifying the North is not, however, generally a consequence of cutting for fuel but of the factors we have already described. There are a few exceptions, as in the Sahel, where some stands of trees have been cut down for fuel, usually by urban fuel merchants. This is a clear example of the conflict between the industrial interests of the city and the wider social concerns of rural society. Older, sustainable practice largely meant the use of the detritus of 'farmed' trees for this purpose, not valuable and productive stock itself. Now, as a consequence of new economic and some climatic pressures, the extensive over-exploitation and destruction of trees has led to extensive degradation of tree-growing land. If per-capita demand continues at the present rate then by the year 2000, with a substantially increased population in the Third World, 2,400 million people will either be unable to get enough wood to supply their minimum energy needs or will be forcing a yet faster rate of land degradation. In 1981 the United Nations, in Nairobi, held a Conference on New and Renewable Sources of Energy. It recommended that the annual rate of tree-planting in developing countries be increased by a factor of five in order to restore the balance and to deal with foreseeable demand. What the conferees undoubtedly had in mind were extensive plantations; however, no funds were found and even this uncertain proposal has remained just a recommendation.

Yet had practice followed the recommendation it is by no means certain that it would have met rural societal needs. These would depend for their satisfaction on sustainability; planting, on the other hand, tends more often to be carried out according to the designs, at best, of the ITTO (in the forests that concern it) or, even less desirably, according to the needs of our development bank officials. Worse still, there have been major plantations which have simply been investment opportunities for Northern companies and therefore bits of private property in which the needs of local people have been rendered irrelevant. Two other, somewhat contradictory, factors need to be taken into account when considering tree-planting. First is the frequent hostility of rural people to plantations that we have already mentioned. The second is the pull of the urban cash economy, which often leads local men to opt for a fast-growing tree cash-crop rather than for the kind of plantation that will meet the wider local needs, which are more commonly taken care of by women.[8]

One of the best known and effective examples of a solution to many of these difficulties is to be seen in the Kenya Woodfuel Development Programme (KWDP). Farming on any scale in Kenya is only possible in the highlands, which are home to roughly 80 per cent of the population and where, as a result, pressures on woodfuel supplies are intense. One of the most densely populated areas is the Kakamega district of the Western Province (roughly 1,000 people per square kilometre), in which the agricultural systems are of the classic integrated and indigenous 'agroforestry' kinds. It is also an area in which trees are most abundant, yet one in which women farmers were finding it increasingly difficult to meet their woodfuel needs, largely because men, who controlled the land, were taking the wood for other purposes. The KWDP was set up in 1983, within a larger programme for woodfuel throughout Kenya, to deal with the problems in the Kakamega and the Kisii districts. Despite variations in farm size, and therefore in circumstance, the KWDP realised that solutions could only lie in woody biomass grown, managed and integrated into local production systems by the farmers themselves. This meant extensive consultations with local people, finding ways in which to overcome the social barriers to communal solutions and taking care to incorporate local needs and priorities. One important aspect of the programme was the provision of seeds, which included those of some exotic as well as familiar species so that the farmers had a wider choice over what they would grow. The cost has been minimal, US$10 million over seven years, and the programme is a model for this kind of approach. Programme workers did not arrive with pre-determined technical solutions to a difficulty, were flexible in their approach and were prepared to adapt to local responses.

Regions under acute stress, like so many of the famine areas of Africa, need a great deal more than plans for agroforestry, though it is important to remind ourselves that, for example, at the height of the Ethiopian famine enough food was available in the country: the hungry simply could not afford it. Famine is more frequently a political and economic than a natural disaster. It has also frequently been observed that famine does not come unexpectedly. A common pattern is for drought to afflict land already degraded as a consequence of external and alien economic pressures; for farmers to reduce their animal stocks by sale in order to buy food that the dry land cannot sufficiently produce; as a last resort, finally to eat or sell their seed stocks and to sell, if they can, their tools. That final stage leaves them unable to survive and famine is full blown.[9] Drought is not new; even the prophet Jeremiah has a cheerful word or two to say about it (Jeremiah 17:8), and people have always lived in dry lands. In the past drought was most commonly due to the failure of the rains; with contemporary patterns of land degradation droughts are now also caused by poor run-offs and low aquifer levels. Ancient strategies for coping with poor rains are well developed among dry-land farmers, but they depend on equally ancient ecological balances and it is these which have been destroyed. Any serious approach to matters of food security for those now starving must be based on enabling the poor to re-establish their own agricultural and forestry patterns.

Preserving existing forests for their people, their biodiversity, their function as carbon sinks, and working towards their sustainable exploitation (whatever that might mean, we may still doubt the legitimacy of those boardroom tables), is unquestionably important. Planting new forest as a crop, in ways not unlike the Nordic approach, could, in certain circumstances and in only a slightly improved world, be a valuable addition to many developing countries' incomes. But the wide encouragement of socially based tree-planting, under the control of local people, would bring enormous benefits of both an economic and an emblematic kind: economic if only because the regeneration of much degraded land and even the improvement of much that has not yet been degraded would then be possible and this, in turn, would lead to fewer breeding grounds for famine; emblematic because it would demonstrate a development for the benefit of the poorest.

Much of this is slowly being accepted in the North, and even the World Bank has heard of phrases like 'social forestry'. However, in common with most other development banks, it still has a penchant for plantations simply because they show quick financial results. Foresters themselves have begun to understand the importance of trees and shrubs in sustaining rural economies and are moving away from their traditional role as protectors and promoters of forests as sources of industrial raw material. Despite their past, they do have considerable expertise and there is a gain in bringing it to bear not just on the question of maintaining the great forests, but also on the problems of re-establishing, where necessary, and helping to develop social forestry. However, the scale on which this is beginning to happen is still very limited and is reflected in the figures for reforestation (see Table 4.2, and the figures for deforestation given in Table 3.2). It is reasonably certain that these figures do not include a great deal of what might be called agroforestry, a significant activity about which bureaucrats can be remarkably blind, but even if this kind of planting were easily quantifiable it is unlikely that it would add hugely to the tiny total.

Table 4.2
Rates of reforestation

Region	*Average annual reforestation 1981–95 (000 ha)*
Africa	
North Africa	101
West Sahelian Africa	10
East Sahelian Africa	37
West Africa	36
Central Africa	3
Tropical Southern Africa	29
Temperate Southern Africa (of which South Africa: 65)	68
Insular Africa	12
Caribbean and the Latin Americas	
Central America and Mexico	32
Caribbean	13
Non-tropical South America	119
Tropical South America	489
Asia	
Temperate and Middle East Asia	972
South Asia	179
Continental South-East Asia	55
Insular South-East Asia	201
Oceania	115
(of which Australia and New Zealand: 105)	

Source: extracted from World Resources Institute, *World Resources 1992–93*, Table 19.1, pp. 286–7, Oxford University Press, New York and Oxford 1992.

Throughout the Third World most farms, homesteads and settled patterns of cultivation are run by women, who make up the majority of the developing world's subsistence farmers. Men may herd cattle and of course there are some who also cultivate, but in the countries with the lowest incomes they are increasingly migrant labourers or workers in part of the cash economy. So it was instructive to turn to the third session of the preparatory committee for the Earth Summit and, in particular, to the report entitled *Conservation and Development of Forests,* a document of some 25,000 words. Almost all of it is about forest appraisal, forest products, what they are worth, their place in world trade and how they might most usefully be used. Section IV deals with the role of women in the use, conservation and sustainable development of forests and takes about 1,200 words. Much of it, while making a ritual assertion that of course anything men can do women can do, if not better then at least as well, explains why they are not allowed to. It is, the report's authors remark, a societal problem in developing countries, and in forestry work of all kinds it is easier to get the views of local men than those of women. Without actually saying so, the section gives the impression that for women to acquire a voice in these matters a change in traditional mores will be needed first and that such a matter is outside the remit of the authors. Exceptions like the Greenbelt Movement in Kenya are described as having more educational than immediately practical value. At no point does this hare-brained document wonder if the reasons why it is so hard to get women to respond are not only social or customary, but also because the wrong questions are being addressed. Neither *Agenda 21* nor the Statement of Principles on forestry is as crass as that preparatory document, partly because throughout all the UNCED documents the principle, if only that, has been accepted that the role of women is important.

Many professional foresters, in spite of considerable changes in their understanding of what is important, have become permanent hazards in the Third World's landscapes. Some of them tend to feel that they have invented the concept of 'agroforestry'. They may well have invented its inelegant name, but the practice is as old as settled cultivation. In the Sahel, for example, traditional farms in many villages are made up of fields, pastures and fallows planted about with trees like the nere, the baobab and the shea-nut; trees capable of producing woodfuel, grazing and food and which promote soil-formation, the preservation of aquifers and so on.

Many of these farms are suffering in the hectic destruction of the Sahelian open-tree forests. If foresters were to combine with agriculturists in programmes designed to help women re-establish and improve the countless patterns of domestic agroforestry throughout the developing world, then some of the right questions would emerge and women would begin to have a voice. Such an approach is wishful thinking in the modern world – balance sheets cannot be cross-sectoral and real, people-based agroforestry is not commercial. We shall return in subsequent chapters to women's issues and land use.

There are, the present authors are inclined to feel, root causes which probably could not be addressed by UNCED and which led to its failure to produce anything other than some fairly anodyne results. This observation is not an emollient for our disappointment, it is merely what must be kept in mind when we consider what ought to have been. Despite affirmations that women have an important role in forestry, that indigenous people have rights, that forests have multiple functions and so on, the emphasis in the Statement of Principles is on management. No-one doubts its importance, but the implication of managerial approaches is that these are always from above. Women, *indigènes*, communities may all be consulted, but the issue of control is evaded and with it the relationship of people both to forests and to extra-forestal tree-cultivation.

When we turn again to *Agenda 21* we find that the first mention of these matters, referred to, not unusually, as 'social forestry', falls in paragraph 14 (d) of Chapter 11, which suggests to 'Governments' such activities as:

> Carrying out revegetation in appropriate mountain areas, highlands, bare lands, degraded farm lands, arid and semi-arid lands and coastal areas for combating desertification and preventing erosion problems and for other protective functions and national programmes for rehabilitation of degraded lands, including community forestry, social forestry, agroforestry and silvipasture, while also taking into account the role of forests as national carbon reservoirs and sinks.

Admirable as these objectives may be, they are listed under a heading called 'Management Related Activities'. Despite the rhetoric of locally managed, people-oriented programmes organised and controlled by the participants, UNCED can only think of their very antithesis, grandiose programmes nationally managed. What was required of the conference, agreements on funding, debt and trade apart, was some examination of the means by which numberless local co-ordinated and unco-ordinated schemes for the regeneration of sustainable 'agroforestry', forestry and the livelihoods around them could be enabled. What we got was business as before, but with socially acceptable trimmings.

There were other elements in both documents devoted to encouraging Third World reforestation as an important contribution to solving the climatic problems caused by Northern industrial practice, protecting genetic resources for purposes as yet unknown, promoting large-scale plantations to increase forest cover (climate worries again) and to meet the world demand for forest products, building proper educational facilities to train future forest managers and technicians. The third preparatory committee in its remarks on 'social forestry' began by suggesting that villagers could be encouraged to plant 'commercial' species and thus help to earn themselves a living. It does talk about other 'utilitarian' uses for trees and brightly remarks that: 'Women and youth [*sic*] have contributed significantly to such efforts in the past.' It suggests two activities for UNCED to back: the promotion and expansion of 'social plantations' and the development and expansion of training for professionals in small-scale forestry.

What we have here is a problem in mathesis; our bureaucrats are mesmerised by the received wisdom of the business world. They are plainly incapable of looking at patterns of development which are not dictated at every moment by the demands of international trade. The idea that sound economic bases might be built, that human survival and happiness might be achieved by moving away from a very narrow financial dogmatism is a little too much for them to grasp. Recovering the basis for the sustainable continued extraction of woodfuel combined with increased food production and the regeneration of degraded land does not look like a promising place for such people to start, for investment would be large and any dividends only too likely to stay with the poor who generated them.

It is the tendency to confuse sustainability and regeneration with commercial practice which is part of UNCED's downfall. Corporate acquisitiveness, worries about climate change, fears of growing Third World populations (the planet, wise Northerners assert with that cant word 'spaceship' hovering in their minds, cannot carry them), worries that somehow otherwise comfortable things in the North can only deteriorate if something is not done – all are at the centre of Northern positions on issues surrounding the 'developing' nations. Forestry focuses these worries because it is an area in which the interests of the poor and those of the rich seem at first sight so clearly to divide. Yet, as in so many other cases, it is not so. Support for a Third World social forestry which restored control and possession to the people would go a long way to relieving a number of Northern environmental fears and might even have a 'sensible' effect on markets.

5

Trade versus Aid[1]

In the preceding chapters we have dealt with the agreements which were actually achieved at Rio. *Agenda 21* differs from them not only because it is not an agreement but because what it covers, no matter how inadequately, bears on almost everything we have to say and we turn to it as the need arises. Now, before we look at some of the other important issues discussed both in the preparatory committees and in the Summit itself, but about which agreements were not reached, it is necessary first to consider the largest single conditioning factor in inequitable development, trade. Constant reference was made during the Summit and in many of its major documents to the need to modify those terms of trade which bear so heavily on the poor. Yet even as pleas of this kind were being made, Northern functionaries elsewhere in the world were busy securing, and even increasing, their trading advantages in the General Agreement on Tariffs and Trade (GATT) negotiations and in those for the North American Free Trade Agreement (NAFTA). We must consider the issue here because no analysis of Southern problems which fails to take it into account makes sense.

The world's major trading agreements dominate all the issues at stake, whether approached from a developmental or an environmental platform. References to trade occur constantly throughout the Rio documents. It is frequently pointed out that trade between North and South could be the engine of development, and behind this lies the unanalysed assumption of some measure of equality between the two 'partners'. Such an assumption becomes absurd when we realise that, between them, and using gross domestic product (GDP) as a measure, the industrialised states produce 85 per cent of each year's global wealth (for GNP comparisons, which effectively give the same picture, see Table 5.1).[2] In addition pleas are made for trading reforms which would favour the nations most in need; in trading and its ancillary activities, like the transfer of technology so often referred to, the 'intellectual' and other 'property rights' of the Northern merchants are to be protected in any concessionary transaction.

Table 5.1
GNP and growth of GNP per capita

Country group		*Average annual growth of GNP per capita (%)*					
	1990 GNP (billion US$)	*1965–73*	*1973–80*	*1980–90*	*1989*	*1990*	*1991**
Low- and middle-income	*3,479*	*4.3*	*2.6*	*1.5*	*0.9*	*0.3*	*–*
Low-income	1,070	2.4	2.7	4.0	2.3	2.4	1.3
Middle-income	2,409	5.3	2.4	0.4	0.4	-0.6	–
Severely indebted	*972*	*5.2*	*2.6*	*-0.3*	*-1.6*	*-3.5*	*-1.2*
Sub-saharan Africa	166	1.6	0.6	-1.1	0.1	-1.6	–
East Asia & Pacific	939	5.1	4.8	6.3	4.0	5.3	–
South Asia	383	1.2	1.8	2.9	2.7	2.6	1.4
Europe	480	–	–	1.0	2.0	-3.7	–
Middle East & North Africa	458	6.8	1.0	-1.5	-1.2	-1.9	–
Latin America & Caribbean	946	4.6	2.3	0.5	-1.1	-1.8	0.7
High-income	*15,998*	*3.7*	*2.1*	*2.4*	*2.7*	*1.5*	*–*
OECD members	*15,672*	*3.7*	*2.1*	*2.5*	*2.7*	*1.6*	*–*
World	*22,173*	*2.8*	*1.3*	*1.4*	*1.6*	*0.5*	*–*

*preliminary data.

Source: extracted from World Bank, *World Development Report 1992*, table A.2 p. 196, Oxford University Press, New York and Oxford 1992.

It is in this last kind of reference to trade, in connection with biotechnology in particular, that Northern alarm bells ring loudest. In paragraphs 2, 3 and 5 of Article 16 of the Convention on Biodiversity we see the emergency services galloping to the rescue:

> Access to and transfer of technology ... to developing countries shall be provided and/or facilitated under fair and most favourable terms ... In the case of technology subject to patents and other intellectual property rights, such access and transfer shall be provided on terms which recognize and are consistent with the adequate and effective protection of intellectual property rights. (paragraph 2)

and:

> Each Contracting Party shall take ... measures ... with the aim that Contracting Parties, in particular those that are developing countries, which provide genetic resources are provided access to and transfer of technology which makes use of those resources, on mutually agreed terms, including technology protected by patents and other intellectual property rights ... in accordance with international law. (paragraph 3)

Further on we find:

> The Contracting Parties, recognizing that patents and other intellectual property rights may have an influence on the implementation of this Convention, shall cooperate in this regard subject to national legislation and international law in order to ensure that such rights are supportive of and do not run counter to its objectives. (paragraph 5)

No serious account is taken anywhere of the fact that allowing new discoveries and innovative techniques to become private property negates the open sharing of information, particularly among scientists, which lies at the heart of much human development.

Agenda 21 carries the same set of messages. Its Chapter 16 deals with issues around biodiversity and biotechnology, in paragraph 7 it points to the need to promote the transfer of technology, but in paragraph 43 it also calls for the strengthening of the capacity of Southern countries to protect intellectual property rights, rights which are, incidentally and in general, the private property of Northern corporations. But it is in Chapters 34 ('Support of and

Promotion of Access to and Transfer of Technology') and 40 ('Information for Decision-making') that the limitations of this form of trade between North and South are more fully spelt out:

> Consideration must be given to the role of patent protection and intellectual property rights along with an examination of their impact on the access to and transfer of environmentally sound technology, in particular to developing countries, as well as to further exploring efficiently the concept of assured access for developing countries to environmentally sound technology in its relation to proprietary rights with a view to developing effective responses to the needs of developing countries in this area. (Chapter 34, paragraph 10)

and in the next paragraph:

> Proprietary technology is available through commercial channels, and international business is an important vehicle for technology transfer ... enhanced access [for developing countries] to environmentally sound technologies should be promoted, facilitated and financed as appropriate, while providing fair incentives to innovators that promote research and development of new environmentally sound technologies. (Chapter 34, paragraph 11)

Chapter 40 implicitly acknowledges that information is a marketable property:

> Existing national and international mechanisms of information processing and exchange ... should be strengthened to ensure effective and equitable availability of information ... subject to ... relevant intellectual property rights (paragraph 19).

The importance of UNCED's emphasis on intellectual property rights lies partly in it being the only really substantive position that the conference took on terms of international trade.

In its second chapter, which enjoys the slightly obscure title of 'International Co-operation to Accelerate Sustainable Development in Developing Countries and Related Domestic Policies', *Agenda 21* makes some very cheering remarks. It points out (paragraph 2) that as long as the terms of trade remain depressed matters are unlikely to improve for the poorer countries of the world, it calls for the liberalisation of trade (paragraphs 3a, 7 and 10) and declares that any 'basis for action ... in promoting sustainable development through trade' must include:

> An open, equitable, secure, non-discriminatory and predictable multilateral trading system that is consistent with the goals of sustainable development and leads to the optimal distribution of global production in accordance with comparative advantage is of benefit to all trading partners. Moreover, improved market access for developing countries' exports in conjunction with sound macroeconomic and environmental policies would have a positive environmental impact and therefore make an important contribution towards sustainable development. (paragraph 5)

It goes on to say that tariff and non-tariff impediments to such a course must be removed (paragraph 7), that protectionist and unilateral policies endanger multilateral trade (paragraph 8) and, oddly, to refer to the hoped for conclusion of the Uruguay Round of the GATT (paragraph 9).

In the North we have lived so long with the benefits of trade agreements that we scarcely notice their existence except on the rare occasions when trade wars between the major blocs erupt into the broadsheet headlines, so it is important to be clear about what is at stake in these demands. The primary instruments of contemporary financial and trade regulation were constructed at the Bretton Woods conference of July 1944 which led to the creation of the International Bank for Reconstruction and Development, now called the World Bank (1945), the IMF (1947) and the GATT (1948).

The GATT is administered through a series of meetings called 'Rounds' and is now subscribed to by 109 countries, who between them account for 90 per cent of world trade. Unlike the World Bank and the IMF, the GATT is, in theory, governed democratically with each member state wielding a single vote. In practice, it is dominated by the industrially developed states since most of the negotiations around which the GATT is defined are formed by private discussions between the US, Japan and the European powers in the Group of Seven (G7) meetings. From time to time the GATT 'rounds' have considered ways of improving trading conditions for the Third World, but little has come of this. Nonetheless there has, for example, been some move to open up Northern markets to Third World textiles by scrapping the notorious Multi-fibre Arrangement (MFA; see below) which, although countenanced by the GATT, lies outside it. Developing countries achieved their own forum in the creation of the United Nations Conference on Trade and Development (UNCTAD, 1964). United States' policy has had its effect on UNCTAD; among its more heavy-handed triumphs was its success in stopping the eighth Round of that conference from meeting in Cuba in 1990. UNCTAD has not succeeded in changing very much and has become less and less important as world trade increasingly falls into the hands of the GATT member states. More and more developing countries are applying for membership of the GATT.

Behind the GATT lies a deceptively simple idea, that all countries will benefit from a free-trade agreement which will encourage its participants to concentrate on those manufactures and services in which they are most efficient or where they have the greatest natural advantage. Doing this will remove the need for the protection of costly or inefficient industries by means of subsidy or tariff barriers, which simultaneously disadvantage foreign competition and are expensive for the taxpayer. Such simplicity is deceptive, because the principle takes no account of major inequalities between nations and by this failure immediately puts poorer countries at a disadvantage. Those countries, attempting to shelter their fragile economies and domestic markets from the onslaught of Northern trading exploitation, find that they are faced with twin attacks from the enforcement of the GATT rules and the use of their indebtedness as a means of compelling trade 'liberalisation' – usually a part of the IMF 'structural adjustment' packages.

The framers of the GATT principles had in mind the depression of the 1930s when, in desperate attempts to improve their position, trading nations imposed an extraordinary series of discriminatory barriers and mutually retaliatory tariffs. To prevent a recurrence of these somewhat self-defeating and short-sighted policies each subscribing nation agrees that every duty it imposes or lifts, under a principle called 'Most Favoured Nation' (MFN), on a particular commodity imported from one country, must be imposed on, or lifted from, similar imports from all other countries. Exceptions may only be made for regions which have established customs unions and areas of free trade, rather like the EC and the North American Free Trade Agreement (NAFTA, covering the US, Canada and Mexico), which may create common external tariffs. However, these are being looked at with increasing distrust by other members of the Agreement as they show every sign of protectionist and discriminatory inclinations.

Another plank in the GATT's free-trade platform is the rule that should any subscribing nation lower its tariffs for the goods of another, then not only should it expect the other party to lower tariffs in return, but these concessions must be offered to all other parties to the Agreement. In order to understand the importance of all this for the developing countries the third major element in the foundation of the Agreement must be borne in mind. This is the principle of 'transparency'. This proposes that in place of tariffs participating nations should employ other restrictive means like import quotas or the recently popular, at least among the most powerful states, 'voluntary restraints'. Safeguards have been agreed in the GATT; for example, it is legitimate for nations to impose duties on imports which are being 'dumped', that is, sold at or below the cost of production – the duty should be equal to the degree of subsidy enjoyed by the exporter.

There are two important areas of world trade which, until the conclusion of the Uruguay Round, are not covered by the GATT. The first is agriculture and this, originally a consequence of the United States' protectionist measures, is an exception which has subsequently been heavily exploited by the European Community and which has led to a major trade war between the two blocs. We shall consider issues to do with trade and agriculture in the next chapter. But important as this is, there is another, fiercer area of restriction on trade, which lies outside the GATT, in the textile industry. This is the MFA, an arrangement formed in 1974 in place of a number of preceding agreements and specifically designed to protect the Northern textile industries from Third World competition.

Because textile technology is, in the main, cheap and production is commonly labour-intensive, it is possible for the industry to be used as a springboard for further industrial development. One of the most striking examples of this may be seen in the development of Ulster's north-east from 1820 to 1914, a process which was, in its crucial early phases, led by the mechanisation of the linen industry.[3] Industrial and economic development are particularly enhanced when the countries concerned are able to add value to their textiles by converting them into manufactured clothing. Northern importing countries, by fixing quotas under the MFA, govern the amounts of worked textiles developing exporting countries may sell them and so control that second stage of production.

Watkins[4] gives a prime and appalling example of the working of the MFA. Pointing out that in 1984 Bangladesh managed to reach a tiny 0.25 per cent of the exports of clothing from the developing to the developed world, he reports that in 1985–6 the USA, France and Britain imposed strict import quotas on Bangladeshi shirts. These were necessary, so the importing states claimed, because the dramatic increase in Bangladeshi production threatened their domestic producers. The clothing industry in Bangladesh was new; it had really only begun around 1980 – the effect of this exclusion was to force the closure of about half of its factories with the loss of 150,000 jobs, mainly those of very vulnerable women.

Southern textile industries could play a vital part in creating an industrial revolution where it is most needed and the undermining of them by Northern protectionism is a gross and cynical disregard of human need, but this is not the most egregious example of these discriminatory practices. As long ago as 1966 the government of Zambia published, as a white paper, an independent report entitled 'Report of the Commission of Enquiry into the Mining Industry', better known as the Brown Report, which, among other things, made the point that the price Zambia received for its copper was set by the cartels of the North. Nothing much has changed since: in 1985 the UN published a survey called *Environmental Aspects of the Activities of Transnational Corporations* in which the point is made again. Some 80 to 90 per cent of the trade in tea, coffee, cocoa, cotton, forest products, tobacco, jute, copper, iron ore and bauxite is, in the case of each of these commodities, in the hands of between three and six transnational corporations (TNCs).[5] In the years 1980–4 developing countries lost some US$55 billion as a consequence of declining commodity prices. These years were the four succeeding the publication of the Brandt Report, which, not least because it pointed to the disastrous consequences for the South of allowing the cartels to fix prices, was met with such indifference.

Among some of the more significant commodities to come from the South are those grouped together under the general name of 'tropical products'. These include, among other things, coffee, tea, fruit and a variety of forest products including palm oil. The trade in all of them is mainly controlled by a small number of TNCs (see Table 5.2). Restrictive practices and price-fixing cartels among TNCs (phenomena to which the GATT seems oblivious) make it rare for the producers to get more than 20 per cent of the market price for these products. Much needed capital could be generated for the South if it were able to export these goods not in their raw state, but as processed manufactures. Northern tariffs make this virtually impossible. Watkins quotes the example of Malaysian palm oil, which in its crude state attracts a tariff of 2 per cent, but if made into margarine before export would be subject to a prohibitive tariff of 25 per cent.[6]

Table 5.2

Concentrations in world commodity trade

Commodity	*%age controlled by between three and six of the largest corporations*
Wheat	85–90
Sugar	60
Coffee	85–90
Cocoa	85
Tea	80
Bananas	70–75
Pineapples	90
Forest products	90
Cotton	85–90
Natural rubber	70–75
Tobacco	85–90
Jute	85–90
Crude oil	75
Copper	80–85
Iron ore	90–85
Tin	75–80
Bauxite	80–85

Source: *UNCTAD Statistical Pocketbook*, United Nations, New York 1984.

In 1990, mainly at the prompting of Japan working with the interests of the developing countries in mind, the states participating in the MFA agreed to its phased demise, but the process is to take ten years. Both the USA and the EC have insisted that during that time they may continue to bring new products into the Arrangement. As all developed countries will first phase out those quotas which are least used, including, possibly, some of the new quotas introduced since 1990, the liberalisation of the textile trade for the developing countries will be a very protracted affair.

Negotiations in the current Uruguay Round governing the workings of the GATT are, at the time of writing, stalled by a failure of the USA and the EC to agree, among other things, upon agricultural policy. If or when agreement is finally reached, this particular round, which has been the longest (it began in 1986) and most complex of them all, will result in a substantial worsening of the economic and trading climate for developing countries. Because the programme of this round has moved the question of Northern limitations to trade out of the main discussions into separate and individual negotiating panels, and because the US, Japan and the EC have combined to push for the rearrangement of trading rules in favour of the TNCs, the interests of developing countries are more or less excluded.

Apart from agriculture, the other main element which prevented the early resolution of the Uruguay Round was the Bush administration's ruthless prosecution of unilateral trade policies in the run-up to the presidential election of 1992. Walter Russell Mead wrote an influential article in *Harper's Magazine* in which he takes a somewhat apocalyptic view of all this.[7] He sees in the US approach to the Uruguay Round and in the creation of the North American Free Trade Agreement a desire to foster a kind of free-trade super-government, unaccountable and overwhelmingly powerful. It 'would be all Bottom-Line: a global corporate utopia in which local citizens are toothless, worker's unions are tame or broken, environmentalists and consumer advocates outflanked'.[8] Thus not only would the poor of the developing countries exist only as a market and as a source of cheap labour, but so, too, would the poor of the developed world. Modern industry, which has shown itself to be increasingly mobile, would move to those regions which offer the fewest restrictions, whether environmental or economic. New rules in the GATT would not only compel developing countries to allow TNCs to move into key and sensitive areas like agriculture and banking, but would also prevent them from constructing any defences which might be construed as restricting trade.

What lies behind Mead's gloomy view is the principal means by which it would be achieved. In December 1991 the secretary-general of the GATT, Arthur Dunkel, offered the negotiators a draft of the Final Act of the Uruguay Round. Among its terms is a requirement that tariffs should be substituted for all other import restrictions. Furthermore, they should, over a period of six years for developed countries and ten years for all but the poorest developing countries, be reduced by 36 per cent, with a minimum of 15 per cent for each group of products. In addition existing access for imports into domestic markets must be preserved and a minimum level of imports equivalent to at least 3 per cent of domestic consumption is to be reached by 1996 and 5 per cent by 1999. The draft Final Act also demands the drastic reduction of subsidies and domestic support prices and their replacement by welfare payments. It is easy to see what effect all this will have on most developing countries, particularly those like India and Nigeria with huge and attractive domestic markets who rely on import quotas to protect their own producers. Faced with the dumping proclivities of the Northern producers of cheap food, their own agricultural policies will be ruined. The effects on the industries of the South will be no less damaging.

Buried deep in the Dunkel draft is the proposal that has, quite properly, most alarmed Mead and which is rarely mentioned in news reports of the Uruguay Round. The GATT is to be superseded by a Multilateral Trade Organization (MTO) which will incorporate its rules, including a new agreement on the trade in intellectual property rights and another on the trade in services. Unlike the GATT, the MTO will have a 'legal personality' which will give it 'in the territories of each of the members such legal capacity, privileges and immunities as may be necessary for the exercise of its functions'.[9] This will mean the subordination of domestic legislation to a supra-national and totally unelected and unaccountable body. Sovereignty will be a dubious quantity and those groups who have fought for and achieved some limited gains in working conditions, health legislation and environmental standards may well see much of what they have won outlawed. So far as the South is concerned, their governments will have little effect on MTO decisions and, instead, are likely to discover that it is just another powerful institution allied to the World Bank and the IMF and one which determines their capacity for action.

Deregulation in support of 'free trade' must be the main purpose of the MTO and it would thus become a charter organisation for the TNCs. These corporations, themselves sitting very loosely indeed to the interests of any nation states, are not simply beyond any normal democratic control but have, by complex systems of interlocking directorships, companies and shareholdings which are almost the structural definition of a TNC, successfully circumvented any serious control by outside shareholders. Power resides entirely in the hands of their ruling boards of directors. They come in all shapes and sizes and usually with a variety of economic and product interests, but one useful indicator of their immense power is the extent to which they control commodity trade (see Table 5.2). We do not need conspiracy theories to account for the undemocratic structure of giant corporations, for democracy is simply irrelevant to their purposes and there is a genuine sense in which the senior personnel in these enterprises would be puzzled by demands for democratic accountability. In the creation of the MTO they will, if they succeed, simply have built an institution suitable to their needs.

That Walter Russell Mead's article should appear in such a prominent journal as *Harper's Magazine* is possibly a sign of the level of opposition to the Dunkel proposal. This opposition is already building in Congress and it is part also of the European, particularly French, antagonism to the GATT agricultural proposals. Opposition from the South is more muted; while it is clear that this Final Act is prospectively a disaster for developing countries since it protects the 1980s unilateralism of the North, they cannot afford to see the complete collapse of the Uruguay Round, which would simply leave unfettered the Northern protectionism the GATT was originally designed to circumscribe.

Since, as we have seen, *Agenda 21* and the UNCED conventions repeatedly, if slightly nervously, refer to the need to safeguard intellectual property rights, we should examine, briefly, what the Uruguay Round is doing about them. That there is a need to transfer technology to the South few would deny, but it is not a simple matter. Much contemporary technology cannot just be transferred like a lump of ore; it needs an industrial environment in which it can be used and developed and, above all, a supporting climate of indigenous research and development. Even in the most advanced of developing countries the latter exists only in fragmented and sectoral form, and it is frequently not to be found at all in the poorer states.

The problem is not simply financial and educational, but also structural. Technological research and development, particularly, for example, in the area of genetic engineering, is ringed about with fences bristling with private patents, mostly in the control of corporations. Many of the new rules suggested in the Dunkel plan, including a proposal for a council governing those issues in intellectual property rights related to trade, are concerned with refining and reinforcing that control. They would harmonise and enforce world patent law and, without concerning themselves with either the original place of invention or the place of production, give absolute protection to patents for twenty years. In the all-important area of genetic engineering, while the rules will restrict the patenting of some animals and plants, all biotechnological processes and genetic materials may be subject to patents. Incidentally the rules will include the adoption of a fifty-year period in copyright; this already exists in the Universal Copyright Convention (ratified by the UK in 1957) and in the Berne Convention of 1886. The effect of the GATT or the MTO adopting the copyright rule will be to enforce it in those developing countries which have, either by consent or by unilateral action, not observed previous conventions.

By and large, developing countries have accepted much of this, partly because they feel that fighting will bring little relief and partly because not to accept it could easily bring yet fiercer trade sanctions down on their heads. A crumb has been cast in their direction: once the Final Act is agreed a period of five years will be given them to implement its provisions; this period is to be ten years for the very poorest. As Mead has pointed out, the adoption of all this was the centre of Bush's policy, not least because it will remove much power from Congress and reinforce the position of the administration. It may not be entirely unreasonable to see something of the sort in the otherwise inexplicable behaviour of the British Conservative government in the 1980s and early 1990s. The net result will be the creation of centralised institutions sitting in a very cosy relationship with the TNCs. Oddly, some opposition to the most radical parts of the Act has come from TNCs – they see the period of grace offered to developing countries as outrageously generous and are demanding a maximum of one year.[10]

That the GATT organisation is sensitive to this kind of criticism is clear, but whether it has any sane answers is less so. In an astonishing little outburst directed at Kevin Watkins and John Vidal, environment correspondent for the *Guardian*, a GATT information officer, one David Woods, wrote a disingenuous letter to Vidal's paper (20 November 1992), long on rhetoric but light in substance, broadly attacking this sort of criticism of the GATT. In it he said that some twenty-eight developing countries 'took to the floor [of the Trade Negotiations Committee] to enthusiastically support the mandate given to Arthur Dunkel'. These countries included 'India, Brazil, most of the rest of Latin America and central America, China, most of the countries of South-East Asia and several from Africa'. He demanded to know whether those of us offering criticism thought that they were all stupid or had, as he put it so elegantly, been 'conned'. We find it difficult to believe that Mr Woods is wholly serious, even he might be expected to know that the economies of developing countries differ one from another. In

South-East Asia a number of countries have, by one means or another, growing and moderately successful economies, but these form only a tiny part of the developing world. China is a special case: it is poor, one-third of the world's people live there and it is only gradually being drawn into the world's trading agreements. The rest of the so-called developing world is trapped in the ways that we have described. Of course they support the GATT because the alternative is worse – they are being offered the amputation of an arm and a leg, in one case with anaesthetic, in the other without. In either case they will be crippled, but the GATT is less painful than no regulation at all.

Throughout the documents from Rio we find references to the need to break down the trade barriers which prevent the South from growing. It is some sort of progress to get the principle acknowledged, but nowhere in the conventions or in *Agenda 21* do we find any commitment from the major trading blocs to practical steps towards its implementation. We need constantly to remember that UNCED was primarily an *environmental* conference, otherwise there is a danger of finding, for example, Chapter 30 of *Agenda 21* merely risible. It is entitled 'Strengthening the Role of Business and Industry' and its first programmatic proposal is to do with the means by which business and industry might more efficiently promote cleaner production methods. To this unquestionably worthy objective it devotes eleven paragraphs before going on to consider how best to promote 'responsible entrepreneurship'. Here, the sanguine reader might suppose, is where some attention would be paid to encouraging responsibility for the needs of the South among the world's largest and most influential entrepreneurs, the TNCs. What this section actually suggests is that business and industry should take a hand in getting the poor to look after their resources a little better and to help them set up small businesses. Dampness in squibs is depressing.

Despite their exclusion from any of the serious business of UNCED, the NGOs meeting in the Global Forum produced among their 'treaties' one on trade and another on TNCs. 'Trade and Sustainable Development' was the title of the former and it describes the negotiations around the North American Free Trade Agreement and the Uruguay Round of the GATT as perpetuating 'the predatory model of development' (paragraph 2). What it effectively proposes, in a lively and fairly sophisticated way, is a sort of Keynesian model of benign liberal capitalism. While its suggestions for action are unexceptionable in theory, it does not really speak to the present case. The 'treaty' identifies unfair trading practices and external debt as the villains in the piece; it urges (paragraph 7) the adoption of the Convention produced by the United Nations' World Intellectual Property Organization (agreed in 1974, it is known as the Paris Convention and embodies preferential treatment for developing countries in both patents and copyright) in place of the GATT proposals; it demands the elimination of all 'trade mechanisms that reduce or restrict the free flow of ideas and technologies necessary for the protection of the environment and health' (paragraph 7); and it calls for a total ban on the international arms trade (paragraph 12).

There is also a paragraph urging that TNCs should be democratically regulated (paragraph 13). This issue was seen to be so important that the NGOs produced a separate 'treaty' proposing mechanisms for the 'democratic control' of TNCs about which, in their very first paragraph, the NGOs took an uncompromising position:

> TNCs are responsible in large part for the global environmental crisis and for many social and economic problems resulting from development. TNCs are main entities in a development process which involves concentration of economic power and production and which leads to social and political inequity and loss of cultural diversity.

What they mean by this is then spelt out in terms of the effects of the activities of TNCs on development, health and the environment.

Solutions take the form of demanding that TNCs respect national sovereignty, that they (part 2, General Principles) observe workers' rights, that they permit 'freedom of information for all citizens, environmental groups, labour unions and governmental agencies' (here they have the problems of dangerous chemicals, emissions and wastes in mind) and generally modify their behaviour in other reasonably obvious ways. To achieve all this the 'treaty' proposed that an electronic-mail 'conference' of NGOs be set up immediately which would produce a newsletter and a list of activities leading to the creation of centres, on each continent, for the gathering and dissemination of information about TNCs. This would 'Support the building up of a countervailing power within countries, involving NGOs, consumer associations, trade unions, citizen's groups, district associations and other grassroots groups' (section entitled

'Action Components', paragraph 1). Perhaps they are right – after all it is the stuff of all good old-fashioned political action – but set in the context of the vastness of international debt, of TNC trading power, of the strength, even in recession, of the major trading blocs, that path may be slow and tortuous and lead to uncertain success.

UNCED and the Global Forum were inevitably trapped by the structures they were criticising, for theirs was an intensely liberal view in a world which has left liberalism behind. The very language of many of the documents betrays their cultural origins – it is sprinkled with words like 'responsibility', 'equity', 'human development', 'dignity' and many such others, all of them concepts moulded by liberal history. They are pitched against the TNCs and the major trading blocs, which find these concepts puzzling. No doubt many of those individuals who hold ruling positions within these institutions would abide by moral codes which would at least be recognisable in liberal society, but their corporations, blocs, whatever, have no use for them. It is pointless to look to trading blocs and their agreements or to TNCs for equitable solutions to world poverty and inequity or to environmental degradation: that is not what they were created for and, by nature, they are unable to provide them. The Global Forum was right in its recognition that countervailing forces are all that can compel the institutions of the global 'free market' to operate within bounds that will allow most people to live decently; it may just have been a little optimistic in supposing that the NGOs could be a focus for their organisation.

6
Agriculture and Land

In order to discuss sustainable agriculture, the wise clerks of UNCED either created or adopted yet another acronym, SARD (it stands for sustainable agriculture and rural development), which they came close to endowing with a will of its own: 'The major objective of SARD is to increase food production in a sustainable way and enhance food security.' SARD is to achieve its objectives by a number of means, which include education and the use of economic incentives and of new technologies. These, in turn, will ensure 'stable supplies of nutritionally adequate food, access to those supplies by vulnerable groups and production for markets' (*Agenda 21*, Chapter 14, paragraph 2). Employment and income-generation will follow, thus reducing poverty; natural resources will be managed and the environment protected. SARD's 'main tools' will be 'policy and agrarian reform, participation, income diversification, land conservation and improved management of inputs' (paragraph 3). We discover, in another sentence in the same paragraph, that SARD's gender politics are a little suspect: 'conserving and rehabilitating the natural resources on lower potential lands in order to maintain sustainable man/land ratios is also necessary.' We shall discuss women and land later in this chapter.

Poverty, rapid population increases and national debt are the universally recognised reasons for the creation, in the developing world, of a massive landless, or near landless, peasantry and a corresponding increase in land degradation and unsustainable agriculture. In the industrial world the causes are other, often just simple profit-taking, but, of course, they tell us, there is always a technical fix for the land an industrial farmer has destroyed. Extremes of unsustainability combined with social and market pressures lead to the horrifying famines we are seeing in and around, for example, the Horn of Africa, even though they are triggered by drought. We made the point in Chapter 2 that famine is rarely, if ever, caused by world (or even regional) absolute shortages of food, but by the collapse of particular, usually farming and pastoral, societies and the lack of incomes they need to buy the food which, for a variety of reasons, they are no longer able to produce for themselves. Perhaps the inability of many people in the industrial world to come to terms with the South's agricultural difficulties springs from the memory of repeated food gluts. The World Resources Institute points out that, in the main, since 1970 the world has been able to produce quite enough food to supply its growing population; it even, in 1990–91, reversed 'a three year decline in word cereal stocks'.[1] However, the WRI also says that it is far from certain that the trend can continue indefinitely and that, anyway, the methods of production used to achieve these levels are unsustainable.

Considerations of this kind lay behind UNCED's views on sustainable agriculture; thus management and policy, not ownership, are the instruments of control proposed in this chapter of *Agenda 21*. Yet intangible social, political and historical elements are involved and these govern the viability of any approach to sustainable agriculture. 'Land', in most of its meanings, has huge resonances for practically everyone. For small farmers, pastoralists and peasant communities throughout the world these have to do with place and family, community and survival. The failure to recognise them is a certain route to unsustainability in agriculture. It is not for nothing that land becomes sacred in many religions and that the ancient agricultural gods are gods of place. Moses had huge problems with them. He also had problems with local versions of that modern scourge, privatisation. Rameses II wanted the Hebrews off his land and into his forced labour battalions in the land of Goshen. He might have succeeded if it had not been for an unfortunate mix-up in an arm of the Red Sea.[2] Much of the Old Testament is made up of national sagas celebrating land.

That loaded phrase to which we have referred before, 'the tragedy of the commons', has a meaning not, we think, intended by Hardin: the tragedy of their appropriation, so far as is physically possible, as private property and their consequent reduction to the status of commodities. The naked and aggressive process of land-grabbing in the North, sometimes, as in many of the clearances in the Scottish Highlands, accomplished by force of arms and the callous use of starvation, was emulated in the colonies. In some places the notion of private and exclusive, personal ownership of land was introduced for the first time. Peasants in many societies, even though they apportioned some land as family plots or farms, saw it not primarily as something any one person owned, but as the means by which the social group lived, as a 'common property resource' (CPR). Nowadays such societies are an endangered species. Where they still exist their own governments have, in general, been won over by what they see as the necessity of private ownership and find the idea of CPRs vaguely threatening. Land 'title' has frequently been used to marginalise women, to grab land and to compel major changes in land use. Some national leaders in the South, like the Marcos family in the Philippines, amply demonstrated their fear of common property with a startling single-mindedness. Northerners are unabashed in their use of real estate terms – Margaret Thatcher remarked: 'No generation has a freehold on this earth, all we have is a life tenancy and a full repairing lease',[3] but then she had a special relationship with the head landlord.

We all depend on land and its uses as surely as we depend on water and air. This trite observation is necessary only because of the degree to which, on the one hand, land, far more than either water or air, is a tradeable commodity and, on the other, the extent to which, as a commodity, it has become part of the international financial network. The edges of that phenomenon are to be seen in the Beaverbrook transaction and in the debt-for-nature swaps in both of which the 'product' rather than the land itself is the commodity. Its centre lies in the latifundias, in the vast cash-crop empires of international corporations and, as in much of sub-Saharan Africa, in élitist private ownership, frequently expatriate, of huge tracts of the best and under-used land. We see in this a fundamental dislocation between the idea of a common resource on which we all depend, an idea to which UNCED and others pay repeated lip-service, and the reality in which all productive agriculture, food security, economic and population patterns are made to fit round that elephantine presence. Tails do wag dogs.

We remarked earlier that romanticism was the ideology by which we moved from husbandry to mastery. That move is exactly what has led us to our present acute concern for the revival of sustainable agriculture. 'Mastery' crept into the English language in the thirteenth century and always seems to have had the dual meaning of both the state of being a master and of commanding a skill. But the idea of authority and control lies behind the word and what better description could there be of the industrial approaches to agriculture that are simultaneously the triumph of, and cause of economic and physical chaos in, modern farming? The destruction of forests and rangelands, particularly in North America, to make way for grain-producing factory-farming is a symbol for the contemporary Northern practice of trying to subdue land and put it to proper use, much as one might shape a lump of iron. In Britain the grubbing up of hedgerows by farming vandals to make way for industrial farming machinery and the consequent poisoning of the land and water supplies with the chemicals invented to replace natural pest-predators and manures is another.

Sustainability in agriculture means farming in ways which do not lead to the gradual destruction of the land by chemicals and by erosion. It is to be achieved by fairly obvious means, but no subject for discussion in development can be conducted without an acronym. In this case it is LEISA, which has an appropriately feminine ring to it. 'Low external input and sustainable agriculture' means making the best use of locally available resources by ensuring that the different elements in the farming system – people, plants, animals, soil, water and climate – work together and in complementary ways. What are widely, if a little awkwardly, known as 'external inputs' are only to be used to make up for deficiencies in the local ecosystem and as a way of enhancing the local biological and physical resources. Care must be taken when using them to recycle as much as possible and to keep bad effects on the environment and on efficiency to a minimum. They should, as it were, be the tools of last resort. The object is to achieve a stable, growing and long-lasting level of production. Farming methods which result in LEISA will obviously vary in the differing economies and ecologies in which it is practised, but combined with land reform the end results will be similar – an agriculture which does not poison the land and the local water resources, and the protection of fragile ecosystems.

This is not a plea for the return to some non-existent and better past, but rather for a change in approach which is a shade less destructive. Everyone is aware of the dangers of contemporary industrial farming, the problem is that it is locked into a huge social and economic complex that renders any alternative difficult to achieve. One major threat to the room for human manoeuvre, which springs from industrial mono-culture, has only recently surfaced. In a fascinating and alarming background paper in an international programme called 'Local Management and Use of Biodiversity', Jaap Hardon and Walter de Boef maintain that plant genetic resources are sharply affected by modern farming.[4] UNCED saw an avenue of progress in providing greater access to global markets for Southern farmers. Hardon and de Boef point to the consequent pressure on these farmers to increase their use of chemical fertilisers and pesticides in order to compete and to be forced by market dictates into growing 'less regionally adapted crops'. Both these elements contribute substantially to the reduction of genetic diversity. The paper's authors also remark that mono-cultures are far more prone to epidemics and plagues of pests and that the best way to overcome these is to fly in the face of modern agricultural practice and plant locally suitable and genetically diverse crops. As an illustration of the dangers of mono-culture they quote the 'southern corn leaf blight epidemic in the United States in 1969–70'. We may see in all this one of several instances in which *Agenda 21* fails to reconcile a serious conflict of interest.

Quite aside from an absolute growth in the world's population, Southern urban settlements are also growing rapidly as a consequence of the flight of the landless or of migrant workers from rural to urban areas. Even so, unlike the North, rural populations in the developing world are not absolutely in decline, they are merely growing more slowly, even in the case of South America where the Southern disparity between rural and urban population figures is probably the greatest (see Table 6.1). From that table it can be seen that during the last thirty years the rate of change from rural to urban population has, on average and throughout the world, increased from 1.3 to 2.8 per cent. While this represents a substantial increase it remains true that for a long time to come at least half the world's population will remain rural. It is, incidentally, the half on which, in one form or another, the rest of us depend for most of our food, so what happens to them and to their land is of the first importance to us all.

In the developed world we have become accustomed to thinking that there are always economies of scale, that if production is amalgamated it is both cheaper and more effective. Large-scale intensive production, particularly of beef and cereals, with 'cheap' technical fixes for the problems, all based on the same philosophy that inspired Henry Ford, has, quite aside from their threat to biodiversity, resulted in some well-known ecological disasters. Some of the most grotesque examples were to be seen in the state-capitalist farms of the former Soviet Union, though there are others nearly as bad in the growing desertification of substantial tracts of US farmland and even of sizeable parts of little East Anglia.

Table 6.1

Rural populations and their rate of change

	Rural population %age of total		*Average annual change (%) 1960-1990*	
	1960	*1990*	*Rural*	*Urban*
World	65.8	54.8	1.3	2.8
Africa (inc. S. Africa and Djibouti	81.7	66.1	2.1	4.9
Central America and Caribbean	67.6	50.7	1.0	3.8
South America	48.3	24.9	0.1	3.6
Asia	78.5	65.6	1.5	3.7
of which China	81.0	66.6	1.2	3.8
India	82.0	73.0	1.8	3.6

Source: extracted from World Resources Institute, *World Resources 1992–93*, Table 17.2, pp. 264–5, Oxford University Press, New York and Oxford 1992.

The world-wide extent of the damage may be seen in Table 6.2 (these are figures for soil degradation produced by human activity of all kinds, but those for the industrialised nations will largely arise from the practices of destructive agribusiness). With a form of agroforestry, a reduction in the use of chemicals, a planned approach to the use of water, an extension of the practice of inter-cropping (poly-culture) and the sage use of genetic engineering, some of these lands, particularly those only 'lightly' degraded, might yet be rescued and still remain as large-scale operations. But it is doubtful whether Ladbrokes would offer very good odds on the likelihood of all this happening.

It is not uncommon for images of poor farmers struggling for a living on marginal land to come instantly to mind when we think of desertification, so we should remind ourselves that rich farmers, not struggling very hard at all, can also degrade land very swiftly.

Table 6.2:
Human induced soil degradation 1945–90

Region	Total degraded (area million hectares)	Degraded area as a percentage of vegetated land
World		
total degraded area	1,964.4	17.0
moderate, severe and extreme	1,215.4	10.5
light	759.0	6.5
Europe		
total degraded area	218.9	23.1
moderate, severe and extreme	158.3	16.7
light	60.6	6.4
Africa		
total degraded area	494.2	22.1
moderate, severe and extreme	320.6	14.4
light	294.5	7.8
Asia		
total degraded area	747.0	19.8
moderate, severe and extreme	452.5	12.0
light	294.5	7.8
Oceania		
total degraded area	102.9	13.1
moderate, severe and extreme	6.2	0.8
light	96.6	12.3
North America		
total degraded area	95.5	5.3
moderate, severe and extreme	78.7	4.4
light	16.8	0.9
Central America and Mexico		
total degraded area	62.8	24.8
moderate, severe and extreme	60.9	24.1
light	1.9	0.7
South America		
total degraded area	243.4	14.0
moderate, severe and extreme	138.5	8.0
light	104.8	6.0

Note: totals rounded.

Source: L.R. Oldeman, V.W.P. van Engelen and J.H.M. Pulles, 'The Extent of Human-Induced Soil Degradation', Annex 5 of L.R. Oldeman, R.T.A. Hakkeling and W.G. Sombroek, *World Map of the Status of Human-Induced Soil Degradation: An Explanatory Note,* rev. 2nd edn, Wageningen 1990, Tables 1–7. Quoted in WRI, *World Resources 1992–93*, p. 112, Oxford University Press, New York and Oxford 1992.

Meat packing in the USA is controlled by three giant companies, Cargill, ConAgra and Iowa Beef Processors – their combined sales in 1987 were in the region of US$47 billion. Roughly one-third of the entire United States is now devoted to cattle-farming and E.G. Vallianatos[5] points out that to produce a little less than 500 grams (1 lb) of beefsteak, the farmer must begin with 2.25 kilograms of grain, 9,500 litres of water and *16 kilograms of eroded top-soil.*[6] Northern agricultural practice may, perhaps, not be the best model to offer the developing world.

Ecological disaster is not the only consequence of the existing forms of large-scale agricultural production; the wreckage of human lives is even worse. The human devastation caused by collectivisation in the USSR, the impoverishment of small farmers during the early decades of this century in the USA and rural unemployment in Britain are all well-known instances. Even the malign effects of the activities of the United Fruit Company and other giant landowning corporations are commonly recognised. But there is one important example which we often miss – the 'Green Revolution'.

Most people are aware that new genetically engineered varieties of cereals (high-yield varieties, or HYVs, now also called modern varieties, or MVs), developed in Mexico and the Philippines, promised huge increases in yield. They were widely introduced throughout the Third World and, planted on large tracts of land diverted from other agricultural purposes, at first showed some surprising results. These, however, were finally seen to be, at least in part, illusory gains: to achieve them called for the much increased use of expensive chemicals (the industrial story behind that is another distasteful example of the ways of Northern producers). Rice harvests, in particular, were difficult to keep because they were more susceptible to moulds than the traditional varieties and all HYVs were vulnerable to a large number of pests and diseases.

However the 50 million extra tonnes of grain which, according to the Green Revolutionaries, were harvested each year amounted to a sizeable increase in the world's grain supplies. That increase did nothing for the poor: even if the kind of grain in question was part of their diet they lacked the money to buy it. In the Philippines for example, where the world-famous Rice Research Bureau has played a major part in increased rice production, two-thirds of the children in Luzon, the main rice-growing area, are malnourished. The landless poor might have benefited a little from an increase in jobs, but their employers who make the real profits from this extra grain are busy developing machinery which is cheaper than human labour. In Thailand and in the Philippines, for instance, new rice-threshing machines are replacing women labourers.

Poor farmers gained nothing. In Mexico, where the main crops among the poor are beans and corn, wealthy wheat-growers, aided by substantial subsidies from the government, ousted many small farmers to make room for their new heavy-yield grain. In India the Green Revolution also began with wheat even though, at the time, only a small proportion of India's agricultural land was given over to that crop. The new seeds depended for their success on careful irrigation and HYV wheat was planted in the most prosperous and well-irrigated areas, when the 17 per cent of Indian farm-land irrigated at the time should have been used for the planting of more diverse and socially useful crops.

The Revolution continued and, by the middle of the 1980s, its apologists were claiming that it was no longer true that its benefits were confined to the larger and wealthier farmers: the poor, too, were planting the new varieties in place of the traditional kinds.[7] Yet in India, for example, the poor, including the poor farmers, continued to get poorer, both in relative and absolute terms, and the process was helped by the HYVs. Poor farmers are not just small versions of the rich, they suffer from a number of special disadvantages. They are unable to buy the chemicals needed by HYVs in quantities large enough to make them cheap; they cannot afford to store their harvests until market prices are sufficiently favourable; and they are not able, either legally or physically, to prevent richer neighbouring farmers from encroaching on what should be common resources. The irrigation on which HYVs depend is expensive and the poor can rarely afford to provide it adequately. As it gradually becomes clearer that successful and sustainable farming, even with the aid of genetic engineering, may return to something more like traditional patterns, the poor will now be unable to afford to make that transition either.

Perhaps the most important thing to come out of the Green Revolution is that the habit of introducing technical solutions to what are essentially social problems has once again been shown to fail. All is not lost, however. Enormous benefits of another kind have accrued from the Revolution. Bank funds, unhappily exposed by, among other things, incautious loans to dubious media barons, are propped up by the massive sums deposited with them by the corporations who have made such handsome profits out of selling petro-chemicals and fertilisers to Third World farmers (see Tables 6.3 and 6.4 for Southern fertiliser and pesticide consumption). In the mid-1980s, the world trade in pesticides alone was worth over US$16 billion[8] and, as our tables suggest, a not unhealthy proportion of that sale was to developing countries. Blessings, like profits, should always be counted.

Small farmers, against these and many other odds, do struggle on but increasingly they are doing so in greater numbers on more and more degraded land. Large mono-cultural farms, economic pressure, dubious title to land in a world where legal and documented title begins to assume greater and greater importance and a host of other problems are either forcing the poor on to marginal farm land or compelling them to unsustainable practices on land which is already fragile. It is well known that the poor have more children as insurance against future destitution and on the principle that the more they have the greater is the chance that some among them will survive. This, combined with the frequently forced movement of the poor away from the most productive land, is producing population pressures which add to the difficulties.

Table 6.3
Fertiliser consumption for cereals, roots and tubers, world-wide

Country group	*1989 (tonnes per hectare)*	*Growth rate 1965–89 (%)*
Low-income:	94	10.3
China and India	138	10.8
Other low-income	39	9.2
Middle-income	69	4.7
Lower-middle-income	60	4.8
Upper-middle-income	82	4.5
High-income:	118	1.5
OECD members	117	1.5
Other	117	1.5
World	97	4.3

Source: extracted from World Bank, *World Development Report 1992*, Table A.7, pp. 202–3, Oxford University Press, New York and Oxford 1992.

Table 6.4
Pesticide use in developing countries, 1975–1989*

Country	*Average annual pesticide use (metric tons of active ingredient)*	
	1975–7	*1982–4*
Africa		
Algeria	16,457	21,400
Burundi	22	59
Egypt	26,970	19,567
Ethiopia	600	993
The Gambia	nd	101
Kenya	935	1,307
Liberia	1,223	310
Libya	2,610	2,017
Madagascar	nd	1,630
Mali	nd	683
Mauritania	11	nd
Mauritius	753	981
Morocco	2,225	3,350
Niger	451	159
Nigeria	nd	4,000
Swaziland	16	nd
Tanzania	2,992†	5,733
Tunisia	nd	1,330
Uganda	nd	23
Zimbabwe	865	207
Central America and Caribbean		
Costa Rica	3,027	3,667
Cuba	7,817	9,567
Dominican Republic	1,961	3,297
El Salvador	1,310	2,838
Guatemala	4,627	5,117
Haiti	156	nd
Honduras	940	859
Jamaica	861	1,420
Mexico	19,148	27,630
Nicaragua	2,943	2,003
Panama	1,542	2,393

Table 6.4 continued

Country	*Average annual pesticide use (metric tons of active ingredient)*	
	1975–7	*1982–4*
South America		
Argentina	7,448	14,313
Bolivia	612	833
Brazil	59,292	46,698
Chile	1,838	1,800
Colombia	19,344	16,100
Ecuador	5,445	3,110
Guyana	705	658
Paraguay	2,957	3,423
Peru	2,370	2,753
Surinam	974	1,720
Uruguay	1,390	1,517
Venezuela	6,923	8,143
Asia		
Afghanistan	1,000	605
Bangladesh	no data	234
Cambodia	1,593	833
China	150,467	159,267
India	52,506	53,087
Indonesia	18,687	16,344
Korea (Dem. People's Rep.)	4,000	nd
Korea (Rep.)	4,675	12,273
Malaysia	nd	9,730
Myanmar	3,721	15,300
Pakistan	2,120	1,856
Philippines	3,547	4,415
Sri Lanka	nd	697
Thailand	13,120	22,289
Viet Nam	1,693	883

Notes:* For many countries there are no data; this table is only an indicator of the size of the industry.
† May not be active ingredients only.
nd: No data available.

Source: extracted from WRI, *World Resources 1992–93*, Table 18.2, pp. 274–5, Oxford University Press, New York and Oxford 1992.

War is another major hazard which increasingly besets the poor. They die, starve and are driven from their land, which is also destroyed, in huge numbers. Mozambique, where South Africa has long waged a destabilising war of intervention, is just one terrible and poignant example of the phenomenon; poignant because left in peace the country has 'a relatively rich but only partly explored and developed resource base, and an overall lack of population pressure on resources with an estimated 18 inhabitants/km^2 in 1987'.[9] Its total population is estimated at 17.5 million. In 1988 two surveys, one by the World Bank and the other by UNEP, both showed that most of its major environmental problems were a consequence of the 'emergency'. Desertification and erosion are very localised phenomena and, in the case of erosion, one of the worst affected areas is in the highlands of the Beira Corridor where so many of the refugees have settled. Similarly deforestation is serious chiefly in urban and peri-urban regions; for example, given peace Mozambique is thought to have a sustainable fuelwood-yield equivalent to almost twice the demand of its inhabitants. Water resources need better management, but are otherwise adequate. Eighty per cent of the country's GDP comes from agriculture and 98 per cent of the cultivated land is made up of family farms, generally of between 1 and 1.5 hectares.

Roughly 200,000 civilians have died as a result of the activities of South Africa's creature, the Mozambique National Resistance (MNR), and well over 200,000 children have been orphaned. In rural areas where war has affected production 2.2 million people are short of food. Another 1.1 million have fled to other areas in their country where they are unable adequately to support themselves. There are also another 1.2 million refugees in neighbouring countries. In a country which so easily could be agriculturally rich, 2.7 million people living in urban areas are now heavily dependent on imported food because of the collapse of the food-marketing system – another casualty of war. It is thought that over 50 per cent of the population is suffering from some form of malnutrition, a figure that should be compared with the 17 per cent in 1983, before South Africa's adventure began.[10] Poverty has increased as Mozambique has had to turn to the World Bank and the IMF for 'aid' in the face of the emergency and has been compelled to adopt structural adjustment programmes. The story differs in each 'developing' country devastated by war, but many of the results are the same.

For those regions still at peace land reform is, of course, the answer to many major agricultural problems, but a world which has made of land a commodity has pre-empted that choice. Once again it is the world of Northern romanticism in which individual rights and property are safeguarded by statutory law, a world which has no time for more communal societies whose patterns of ownership depend on customary law. Statutes and the model of justice on which they are formed can, by their nature, only be concerned with legal title and not with any moral entitlement, which is more likely to be recognised in customary law. The idea that peasants, small farmers, are entitled to land, particularly when land is incorporated into their societal relationships by marriage, common resources, shared skills and so on, is one that lodges uneasily in Northern minds. So it is that land reform is often seen as a dubious form of charity, not as a prerequisite for justice and, indeed, for development.

In 1988 the FAO pointed out that landlessness 'is emerging as possibly the single largest agrarian problem in the world'.[11] Three-quarters of the world's people live in the Third World and over half of those are in rural areas. No matter what the rate of increase in city populations may be, and it is already very rapid, there will still be a vast population of small farmers and landless peasants. Many small farmers, particularly women, are forced off their patches of land because of dubious title, or because the land has become degraded to an unworkable point, or because of ill-health or war. Often their only hope of survival is to become landless labourers on the farms of the wealthy or to become sharecroppers. In the case of men, in some countries becoming migrant industrial workers is another strategy for survival.

We have models for all this in the Northern past, but they are misleading. In the first place it is a matter of scale, the size of the phenomenon in the Third World defeats the imagination; in the second, the destruction of the peasantry in the North went hand in hand with the creation of an industrial economy which eventually led to the world's existing power relations. There is little evidence that any such transformation, no matter how brutally it might emulate Northern history, is being permitted in the South except in a very few instances under the tutelage of Northern interests.

Faced with horrifying famines, particularly in the continent of Africa, the press and television have invented 'donor fatigue'. In their blundering fashion they have come within sight, even though they have missed it, of the real problem. Huge numbers of ordinary people are still compassionate and generous, but it is beginning to dawn on them that although giving food, cash, medical supplies, blankets or whatever may save some lives, in the end it changes nothing. There is a well-known dilemma here: emergency aid does save lives, and it is important that this should continue, but equally often it has the effect of depressing local food prices and so forcing yet more marginal farmers into destitution. Whatever the short-term responses to it may be, it is important to realise that famine is the last stage in a lengthy and continuing process in which formidable numbers of people in the Third World are involved. Emergency aid, as many of the more respectable Northern NGOs have pointed out, will always fail in its object unless it is accompanied by a host of other measures addressed to much earlier moments in the process towards famine.[12]

Large estates, latifundias, which are inefficiently run, agribusiness producing mono-cultural cash-crops, are all common enough in the South, but the huge majority of farming people there, with varying rights to parcels of land of equally varying sizes, are in effect peasant farmers. Many are subsistence farmers; others farm for the market, often in association with neighbouring agribusiness; many more operate a mix of subsistence- and market-farming. What is central to peasant communities is the issue of land control. To use the word 'ownership' is to muddy the waters because it has such peculiar connotations in the North which are purely to do with European medieval legal concepts; were it not for that, then it would be the word we need. Whatever the correct expression may be, the one question above all which affects the notion of sustainable agriculture is that of the 'legal' relationship of the peasant to her land. *Agenda 21* is our touchstone, the 'commitment' made by the world's governments to progress – it is therefore instructive to see what it has to offer on the question.

Much of that document is chiefly concerned with land management and policies. There is nothing particularly wrong with this except for the general uselessness of putting carts in front of horses. There is a patrician air about large parts of *Agenda 21* which leaves the reader with the uncomfortable suspicion that some, at least, of those who drafted it really felt that international, national and regional bodies are the places where the power to do things resides and that is up to these organisations to create the right ambience for their subjects. They tend to forget, if indeed they ever knew it, John le Carré's apposite remark: 'A committee is an animal with four back legs'.[13] A land reform based on the needs and abilities of Southern farmers is scarcely discussed.

Yet there are, after all, plenty of miserable object lessons around. We have only to think of Zimbabwe, a country in which all the best agricultural land was appropriated by white settlers in its days as a colony and much of it used under-productively in cash-cropping. A huge landless and largely unemployed peasantry was created and settled in townships, without adequate infrastructures, to a life on the bread line. With independence we might have expected to see some reform. But fear of South African intervention, added to by the spectacle of that country's performance in Mozambique, fear of reprisals from international finance and a large measure of internal dissension, all mean that the Zimbabwean government, despite its citizens' demands, has hardly moved on the question.

Other voices can distantly be heard in the text of *Agenda 21* through the overwhelming chatter about the issues of, and surrounding, land management. Chapter 3 on 'Combating Poverty' in paragraphs 5 (c) and 9 (o) refers to the needs of the landless; Chapter 7 on 'Promoting Sustainable Human Development' mentions, in paragraph 9 (b), after having first talked about the need for sustainable management, the farmers' need for protection against unfair eviction, the need to promote fair 'access' to land and so on. Chapters 12 on 'Managing Fragile Ecosystems: Combating Desertification and Drought', 14 on 'Promoting Sustainable Agriculture and Rural Development' and 32 on 'Strengthening the Role of Farmers' have, as indeed they should, the best passages. Paragraph 24 (a) of Chapter 12 calls for the establishment of 'mechanisms to ensure that land users, particularly women, are the main actors in implementing land use'. It elaborates this view with a demand, in paragraph 28 (c), that 'Particular attention should be given to protecting the property rights of women and pastoral and nomadic groups.' Chapter 14 asks for 'policies to influence land tenure and property rights positively with due recognition of the minimum size of land-holding required to maintain production and check further fragmentation' (paragraph 9 [c]), and paragraph 14 (b) wants 'legal measures to promote access of women to land (*sic*) and remove biases in their involvement in rural development'.

Chapter 32 has several of the same demands but adds to them, in its paragraph 5, the observation:

> The key to the successful implementation of these programmes lies in the motivation and attitudes of individual farmers and government policies that would provide incentives to farmers to manage their natural resources efficiently and in a sustainable way. Farmers, particularly women, face a high degree of economic, legal and institutional uncertainties when investing in their land and other resources. The decentralization of decision-making towards local and community organizations is the key in changing people's behaviour and implementing sustainable farming strategies.

That bears all the marks of a nail struck firmly on the head but, alas, these voices are rare and cry alone in the wilderness of technocratic eco-fixes.

There are two inseparable starting points in any fight for sustainable agriculture: the rights of women and land reform. We have already remarked that most of the world's subsistence farmers are women: 80 per cent of the food for sub-Saharan Africa is produced by them; in Asia the proportion is between 50 and 60 per cent; in North Africa, the Middle East and Latin America it is about 30 per cent.[14] Societies do exist in which women have, at least in theory, equal rights in land, but they are few and even then those rights are often severely compromised by male interests. Because women's rights are a major social, political and economic issue, it is customary for most Northern aid workers to skirt round it – they see it as something to be dealt with only by the societies involved. By standing aside and simply getting on with their 'programmes', they collaborate, no matter how unwillingly, in preserving the status quo. Because so many of the world's farmers are women, because for a host of historical reasons they are often outside the processes of the urban, 'free-market', cash economy, no serious restoration of land resources to the people is possible without laying the basis for securing their rights.

The literature, that somewhat uncertain 'literature of hope', is stuffed with examples illustrating the connection between the rights of women and sustainable farming practice. Perhaps one of the most important of them is seen in the efforts to halt the desertification in the Sahel caused by poor farmers who are repeatedly forced into the over-exploitation of marginal land, particularly in the fragile areas of dry land. In most of these projects, organised by NGOs or the state concerned, women make up more than 50 per cent of the workforce and in some instances the proportion can reach 90 per cent or more. They are commonly allotted the heaviest and most menial tasks, but despite this many of them are acquiring new skills in surveying, market-gardening and arboriculture. In general the women work either as volunteers or are given food in return for work – any paid jobs are, with a few notable exceptions, usually taken by the men.

Skills have been gained and by working together, particularly on desegregated sites, women have come to recognise their own strengths. This has meant a new confidence in their own capacity to organise and to work for change. Although this is a very important beginning and it is reasonable to expect that, in time, matters will change, inequalities remain. Women who are employed in arduous work on these projects have, in addition, another full domestic workload. Only 6 per cent of the land reclaimed in the Sahel, largely by their labour, has been allocated to them. Though they may only be paid in kind, the employment of women has encouraged even more men to leave the villages to become migrant industrial workers, secure in the knowledge that their families are provided for by the women's extra work.[15] The allocation of land is still entirely a male prerogative, but it is a not wholly unreasonable hope that this position, too, will be eroded by the sheer weight of women's numbers and their continued and increasingly organised presence on the land.

Examples exist in other cultures, among the most interesting of which is the land reclamation scheme launched in 1980 by the women of Bankura in West Bengal.[16] This district had once been covered in forest, which had provided the habitat and resources for an ancient and traditional community. It was destroyed as entrepreneurs realised the cash value of the forest products, including the trees themselves, in markets outside the communities. The inevitable degradation of the land drove even the women to working, often as bonded labourers and in dreadful conditions, for large and usually distant landowners. Periodically released from their bondage they would often return to find that the houses they had left had collapsed and that life had to be started all over again. In a moment of revolt against all this a group of women persuaded some owners who could not afford to improve their degraded and useless land to give them title and they persuaded local government officials to help with minimal finance. It took them three years to reclaim the land, which is now thick with trees and fertile. Even more importantly, only seven years after they began,

1,500 women in the immediate area, some with small plots of land to donate, others landless, formed similar collectives, many of them extending their means of generating income. The movement is spreading fast and there is no reason, other than male prejudice and bureaucratic fears of democratic collectives, why it should not go a long way towards answering the desperate questions posed by land degradation and world poverty.

The present authors' jaundiced view of 'the literature of hope' springs more from its function in obscuring questions among Northerners than from any reservation about the quality of the experiments it describes. We are also well aware of how pathetically small these movements in the direction of sustainability as yet are. One of the commonest ways politicians, bureaucrats and some of the patriarchs of the more conservative NGOs have of dismissing them is to point to that smallness. However, we must insist that the failure adequately to try these methods or the failure yet to overcome the largely male opposition to them is no argument for supposing that they cannot work. Nor can an obsession with the centralising free-market system be a reason for decreeing them to be either a side issue or irrelevant. They are at the centre of any solution to rural poverty, any future food security, any environmental preservation and, in the end, to any worries about the effects of a drastically increased world population. In this context another very important aspect of women's rights to land lies in their unique recognition of the connection between land degradation and untenable domestic workloads. It is probably this more than any other factor which has led women to adopt the most productive mixed-farming practices possible in each environment – to develop, in other words, sustainable farming methods. Establishing their rights, together with making land at present kept either idle or inefficiently productive by the world-wide *latifundista* styles of management, would, even in the fairly short run be every bit (if not more) important than wonder-science's techno-fixes.

These are, of course, counsels of perfection and if all that stood against them were venal politicians, machismo and some tradition then women could probably be left to fight their own battles successfully, but it is not as simple as that. Agricultural production is very big business: in the years 1970–89 world cereal production rose from circa 1.2 billion tonnes to circa 1.8 billion tonnes; in the centrally planned economies of the developing world food production generally doubled in the same period; production in Africa increased by 40 per cent and in the Near East by 60–80 per cent. Most of these increases have been a consequence of increased yields rather than any extension of the areas farmed. Livestock populations are some 18 per cent higher, and between 1969 and 1989 fish catches increased by 67 per cent.[17] Much of this increase took place in developing countries, which, until the early 1980s, had been a valuable market for Northern food surpluses.

Increases in Southern production should have been the golden opportunity for Northern states to pull out of much intensive farming and, by the careful use of taxation, to encourage a range of environmentally sound practices; enough was known about global environmental problems at the time to support such wisdom. Instead Northern governments and agribusinesses, with smaller farmers dragged in their wake, intensified the production process which led to the massive food surpluses of the mid-1980s. Subsidies and other devices, paid for out of taxation, largely protected the producers, but the sums were colossal: from 1982 to 1985 the United States' government farm expenditure increased from US$5 billion to $22 billion; the Common Agricultural Policy involved the EC in expenditure which increased from $10 billion in 1984 to $20 billion in 1986 and, an OECD report noted, almost half the income of European farmers came from subsidies.[18]

Subsidies of this kind were designed to achieve, at almost any cost, food security for those states giving them and a dominant position in the world market. Nigel Harris, of the Development Planning Unit of University College London, made the comment, only half in jest, that they should really be seen as part of the defence budget.[19] As well as depressing world food prices they have resulted in grotesque surpluses in the North and one other effect of this has been to marginalise many small Northern farmers as well as those in the South: 75 per cent of all European agricultural produce now comes from 25 per cent of its farms. This is a further cartelisation which will also rebound on the South. There are less than edifying stories of British farmers being paid to destroy crops rather than adding to the surpluses. In 1980 an attempt was made by the EEC to sell 20,900 tonnes of surplus butter to the Soviet Union which, because of the gap between the EEC buying price and the price to be charged to the USSR, would call for yet a further subsidy of about US$1,900 per tonne. In Britain little packets of butter were handed out to old age pensioners with their pensions.[20] Northern states were quick to see the advantage of using some of these food surpluses as non-emergency aid to the Third World, for it was far cheaper to export them than to store them.

Table 6.5
World commodity indexes for all agriculture and for cereals, price indexes for beef, sugar and wheat, 1975–1989

	Commodity indexes (based on constant prices, 1979–81 = 100)										
	1975	*1980*	*1981*	*1982*	*1983*	*1984*	*1985*	*1986*	*1987*	*1988*	*1989*
All agriculture	100	104	91	81	89	94	81	71	59	63	61
Cereals	142	101	107	79	87	85	74	55	47	59	64
	Commodity prices (in constant 1980 US$ per unit of measure)										
Beef	2.11	2.76	2.46	2.41	2.53	2.40	2.25	1.85	1.93	1.89	1.95
Sugar	0.72	0.63	0.37	0.19	0.19	0.12	0.09	0.12	0.12	0.17	0.21
Wheat	288.7	190.8	195.4	168.0	175.5	174.3	180.7	141.6	108.2	134.7	153.0

Source: WRI, *World Resources 1992–93*, extracted from Table 14.4, p. 242, Oxford University Press, New York and Oxford 1992.

What the North had produced was a world glut of very cheap food and it was marketed at prices well below those of production (see Table 6.5). From a mean taken in 1975 there was an absolute decline in agricultural prices of 39 per cent by 1989. Cereals in general dropped by 55 per cent, and wheat in particular by 47 per cent. Beef went down by 7.5 per cent and sugar by a massive 71 per cent. One immediate consequence of all this has been the virtual destruction of any efficient Southern producer who had to work without the massive subsidies that have so depressed world food prices. By the period 1980–84 OECD states had increased their share of the world beef market from 40 per cent to 62 per cent; their sugar trade rose from 17 per cent to 28 per cent against a drop for the South from 73 per cent to 67 per cent.[21] It is this last change which is the most notorious: the EC and the USA have, between them, devastated the developing world's sugar industries. By continuing massively to subsidise sugar-beet production, by fiercely reducing quotas for imported cane sugar and by dumping huge supplies on to the world sugar market, these two blocs have destroyed the International Sugar Agreement, produced what seems to be an endless slump in world prices and imposed yet greater hardship on the already poverty-stricken sugar workers throughout the South. It is an irony that we have simultaneously to recognise the devastating blow to some Southern economies in this story and to observe that sugar-cane, in the first place, is a classic example of colonial mono-culture introduced to satisfy Northern markets and one which renders other forms of agriculture marginal.

The grotesque heavyweight agricultural title fight between the USA and the EC is causing terrible damage to Southern economies as the contestants continue to slug it out with idiotic over-production and subsidies likely to beggar all their taxpayers. In sub-Saharan Africa one-fifth of local consumption is supplied by cheap imports which have to be paid for with one-fifth of the region's foreign exchange earnings. The value of Thailand's rice exports fell, in the late 1980s, from US$230 per tonne to $170 per tonne. Up until the 1980s half of Argentina's export earnings came from its production of cereals and vegetable-oil seeds; during the 1980s, as a consequence of US and EC policies, these earnings fell by 40 per cent.

Watkins[22] gives other examples. During 1986, the battle between the USA and the EC resulted in both selling wheat, subsidised to the tune of US$100 per tonne, in countries like Mali and Burkina Faso for as little as $60 per ton. Locally produced cereals like sorghum and millet cost three times as much as that. Peasant farmers' incomes fell, acute hardship followed and farms, no longer producing enough cash to allow for normal investment, were degraded. Growers of root crops suffered in much the same way. Pastoral farmers do not escape; in 1990 the International Fund for Agricultural Development (IFAD) reported that many of its agricultural developmental projects in the Sahel were being undermined in much the same way by imports of cheap EC beef into the region.

Some Southern governments encouraged this process because, for a time, it meant cheap food could be provided for their politically volatile urban centres, but it had a price. Subsidised food imports may be cheap, but they are not free and had to be paid for with scarce hard currency and this added to the already substantial Southern balance of payments deficits and consequent indebtedness. Far worse than that, however, was the creation of a market climate of food-import dependency which seriously undermined Southern food security. Cheap imports, in ruining local farmers, destroyed substantial parts of the local productive base. They also hastened the dash towards mono-cultural cash-cropping for Northern commodity markets and the sugar story shows how fragile that recourse can be. Not only do the Northern subsidies reduce Southern food security and increase Southern poverty, they also tighten the Northern grip on the agricultural commodity market.

If we turn from this part of the problem to the effect of the Uruguay Round of the GATT on agriculture in the developing countries then we come upon an even bleaker picture. Import quotas, tariffs and subsidies should be in place in all those Southern states with an agriculture to protect. However, the USA and the EC have generously agreed that all Southern countries will share in the liberalisation of trade and that means that they may not use these protective devices even though, hitherto, agriculture has fallen outside the GATT rules. There are exceptions; the devices may be used provided that they in no way interfere with trade and that they do not keep domestic prices higher than those known as 'free-at-frontier', that is the price at which the exporter would sell where restrictions do not apply. This rule, which, in the real world, is derisory, because such prices are dumping prices, means that virtually all peasant farmers in the developing world will face competition from the massively subsidised producers in the USA and the EC. Some states, India and Nigeria among them, have used import controls and price-supporting mechanisms to protect their farmers, but the USA has already made clear its intention of having these forbidden under the GATT rules. The North is allowing the GATT to encompass agriculture as it suits it to do so.

From time to time UNCED mentioned that it would be a good thing if something were done about all this, but it did not do anything so lacking in taste as to suggest what. This was left to the commonality in the Global Forum who came up with a spirited set of responses. They began uncompromisingly in paragraph 8 of their document on sustainable agriculture: 'The so-called "liberalization" of agricultural trade, as proposed in GATT, will strengthen the monopoly of the present agro-economic system and uniformity of production systems.' In their programmatic section they continued:

> WE NGOs & SOCIAL MOVEMENTS PLEDGE TO:
> Support efforts and interaction between people's movements, women's groups, youth, indigenous peoples, local communities and peasant, family and small farmers' organizations [to] enhance and maintain intact sustainable farming systems, to restore degraded agroecological and cultural systems, to accelerate development and implementation of sustainable agriculture practices [paragraph 19];
>
> Emphasize the need for people's empowerment, especially the active participation of women ... to obtain access to land tenure, credit, training opportunities and education ... Pressure governments, agricultural research institutes and rural extension agencies to include or increase participation of peasants, family and small farmers and rural residents in the decision-making process, and to base research and funding decisions on direct consultation with and consideration of the needs and priorities identified by farmers [paragraph 25];
>
> Increase sustainable agriculture production in urban, peri-urban and rural areas at the grassroots level with an emphasis on alleviating poverty and improving regional food supply, small scale production and self-sufficiency [paragraph 29];
>
> Advocate ... democratic and equitable distribution of land wealth through the principles of agrarian reform based on the control of the land workers [paragraph 30];
>
> Campaign for international negotiations on agricultural trade practices, notably in GATT, which encourage and support policies on sustainable agriculture, ensuring that the issue of food security and the health and nutrition of all people be given top priority, with emphasis on poor people [paragraph 37];

These extracts are not a summary of the Global Forum's 'treaty', and the remainder makes a number of serious points which speak to our concerns. Even more, it sets out, programmatically, something of a social agenda for the South and all those who work on its behalf. It is good that the Forum raised a standard about which others may rally, the beginnings of a genuine *Agenda 21*. This agenda is to be built around a number of major policies. Market pressures which push farmers towards unsustainable industrial practices must, by a system of tariffs, import quotas, and price-support mechanisms, be modified. That means unrelenting pressure on the existing trade blocs, and any agreements that they may produce, to make special provision for the South. Land reforms designed to release much of the under- or ill-used giant estates for peasant occupation must be undertaken. Women's land rights, credit unions and trading co-operatives, for reasons to which we have already pointed, must be fought for. All this is to be supported, not dominated or led, by extensive programmes of research into land regeneration; new, as well as old, sustainable agricultural methods; and *market* as well as land management. The relationships between rural and urban communities are symbiotic. Part of urban regeneration must be massive investment in job creation, but this must also be extended to include many of the landless, not all of whom could necessarily find, even in a sustainable world, enough work in farming.

Our list is not exhaustive and it is not the purpose of this book to suggest means by which such programmes might be achieved, except to say that they will call for immense shifts in the political and economic priorities of the North. What is clear is that no matter how well intentioned and even successful local instances of revived sustainability may be, they can be no more than beacons, exemplars, until the larger reforms begin to emerge.

7

Water, Salt and Fresh

Our planet is a very wet place and those of us who inhabit the Celtic offshore archipelago in the North Atlantic are usually very well aware of it. About 70 per cent of the world's surface is covered by sea water; the Pacific Ocean alone accounts for 32 per cent. Probably the oddest thing about our relationship with the wet two-thirds of the world is that so little is known about the oceans. Myths are common – from Oedipus's encounter with Scylla and Charibdis through the Flying Dutchman, to those wonderful triangles where ships and aeroplanes disappear – but research is thin. Currents, marine biota, the ocean's geomorphology and the relationship between sea and climate are only some of the major areas in which little more than broad patterns are recognized. That 60 per cent of the world's population lives within 100 kilometres of the coasts probably tells us more about geography, the history of transport and politics than about any atavistic longing for the sea, but it becomes important when we look at the effect of this concentration on the so little understood marine environment. In the Third World the human pressure on the coasts takes place for reasons very like those compelling desertification on fragile dry lands – people fly unsustainably to cities, many of which are coastal, because they have been driven there by poverty.

Governments are schizoid about the seas. On the one hand they treat them, quite correctly, as common resources and, on the other, they do their utmost to control them. The USA during its long Cold War on the Soviet Union made of the Pacific an 'American Lake' in which, with very small exceptions, US military control was complete. Similarly the Atlantic and even the Indian Oceans are policed by the navies of the United States and its allies. The fate of the Antarctic Ocean hangs in the balance as a number of powerful states, coveting both its resources and those of its continent, jockey uneasily for position. Economic Exclusion Zones (EEZs) of 200 nautical miles (370 kilometres) are now part of the law of the sea and, for instance, the smouldering UK–Irish quarrel about Rockall has to do with these zones. In general most states strong enough to have an effective view give the impression that although they recognise that the seas are mainly 'common', they feel that in a decently ordered world they would unquestionably be private and belong to them.

All coastal states with a splendid disregard of what little is known about the seas and their currents have used them as dumps, particularly for sewage, with the result that, in varying degrees, there are now no unpolluted coasts. The early 1990s cholera epidemic in Latin America was probably partly spread by means of sewage-contaminated shell-fish. Polluting and polluted substances borne away from coastal areas by currents have combined with deep-sea dumping to pollute both the sea bed and places far away from the source of the wastes. Polychlorinated biphenyls (PCBs) have even been found embedded in the Arctic ice; toxic chemicals and radioactive wastes can drift about the world. Oil is spilt, particularly from the ageing fleet of very large crude carriers (VLCCs), in quantities that defy belief. It is estimated that in the Mediterranean Sea alone between 500,000 and 1 million tonnes of oil and related products are spilt or dumped every year.[1] Plastic packaging and abandoned plastic fishing nets have become a major marine hazard, particularly in the Pacific. Disturbing reports of nuclear weapons carelessly mislaid at sea appear from time to time, in the newspapers. The catalogue is endless.

We have already mentioned (in Chapter 3) the continued destruction of delicate coastal ecosystems and the consequent reduction of biodiversity. Much of that destruction is caused by the increased concentration of coastal, or near coastal, populations particularly, but by no means exclusively, in the Third World. If any of the assumptions used by Krause prove to be correct, then two even greater difficulties will strike us. Rising seas will finish destroying much of what people have so far left alone and coastal populations will be increasingly threatened. That this danger is very present in the minds of Southern leaders is well illustrated by Bikenibeu Paeniu, prime minister of Tuvalu, announcing the need to threaten the functionaries of the British Foreign Office with a rupture in relations if Britain continues to ignore Tuvalu's need for help in maintaining sea defences.[2] The developed world will, no doubt, spend huge sums on its own sea defences or, in extreme cases, on evacuation, but where will the South find the resources to cope with yet greater ecological catastrophe?

Many of the problems in this brief but dismal catalogue are susceptible of at least partial solutions. It may be difficult to stop dumping at sea, to reach the necessary agreements, to knock the heads of sectorally divided bureaucrats together, but given the political will it seems at least possible, and a substantial start has been made, particularly with the UNEP Regional Seas Programmes. Rising seas, however, are altogether another matter. Most of the South's industrial assets, together with some of its largest food-producing regions, lie at, or even below, present sea levels; among them are the vast deltas of the Nile, the Mekong and the Yangtze. Cities in danger include Calcutta, Lagos, Rio de Janeiro, Cairo and Bangkok, but there are countless others. Martin Ince, in his admirable small book on sea rise,[3] assembled Table 7.1, which shows clearly that even Surinam, the richest of the ten, is unlikely to have the resources to deal with successions of major disasters. Even though total inundation, except in very unusual cases, is probably not to be expected, frequent flooding, sea surges, salinisation and coastal erosion will become common. We have only to think of the dreadful loss of life and environmental damage caused by the repeated flooding of Bangladesh to get some inkling of what this might mean.

In 1990 UNEP published an excellent report entitled *The State of the Marine Environment*, which was a first and important step in understanding the scale of marine pollution and its causes. It describes the growing urbanisation of coastlines. The most dramatic example is probably the Mediterranean Sea, where there is now virtually no unbuilt-upon coast; the population along its shores is expected to rise to 200 million by the end of the twentieth century. In other parts of the world coastal wetlands of all kinds are also disappearing, partly as a consequence of population pressures but also to satisfy Northern demand; more mangrove plantations are destroyed each year in order to supply the Japanese wood-pulp industry than are degraded by excess numbers of people. As we pointed out in Chapter 3, the fact that wetlands and mangroves are vanishing (see Table 3.1) will affect the world's fish stocks if only because of the percentage of commercially caught fish for which they are the spawning grounds.

Table 7.1
Ten countries most vulnerable to sea-level rise

Countries	*Population (millions)*	*Per capita income (US$)*
Bangladesh	114.7	160
Egypt	54.8	710
The Gambia	0.8	220
Indonesia	184.6	450
Maldives	0.2	300
Mozambique	15.2	150
Pakistan	110.4	350
Senegal	5.2	510
Surinam	0.4	2,360
Thailand	55.6	840

Source: Martin Ince, *The Rising Seas*, Earthscan Publications Ltd, London 1990.

The report also describes the problem of waste pollution of the seas by shore-dwellers and we may take the South Asian sub-continent as an example. Calcutta alone dumps 400 million tonnes of raw sewage and municipal wastes into the Hooghly, which flows into the Bay of Bengal where it joins the detritus of Bangladesh. Karachi pumps 175 million tonnes of untreated sewage into the Arabian Sea every year. India as a whole dumps about 34 million tonnes of solid wastes into its coastal waters. The Indus carries vast amounts of pesticides to the sea. Small attempts are being made to deal with all this; thus the World Bank is lending the municipality of Bombay some money for the construction of sewage works, but it is unclear whether they will be large or effective enough to cope with the increased demand by the time they become operative within the next ten years.[4]

None of this is special to South Asia. An advisory panel set up by the United Nations and called the 'Group of Experts on the Scientific Aspects of Marine Pollution' (GESAMP) reported, in 1990, that there was serious and world-wide spoilage of coastal areas by atmospheric pollution and, most importantly, by sewage and agricultural chemical run-off. Huge increases, varying from 50 to 200 per cent, were detected in the amounts of nitrogen and phosphorus entering the seas. These increases were caused by an estimated annual discharge, over and above 'natural' discharges, of between

7 million and 35 million tonnes of dissolved nitrogen and between 0.6 million and 3.75 million tonnes of dissolved phosphorus. Most of these nutrients stay on the continental shelves, where they frequently give rise to destructive, and sometimes dangerous, algal blooms. Their concentration around the coasts is such that 'only 0.44 million metric tons of nitrogen and 0.3 million metric tons of phosphorus are added to suspended sediments in the deep ocean each year'.[5] The depletion of oxygen in the coastal waters caused by this over-fertilisation is one problem, but there are others. Pathogenic bacteria and viruses, heavy metals and a cocktail of dangerous chemicals and radioactive materials all come as part of the package.

UNEP's report on the marine environment is mainly concerned, reasonably enough, with coasts and it is not really surprising to discover that patterns of pollution reflect the economic status of the polluters. Developed countries destroy their coasts and surrounding seas largely with industrial, nuclear, agricultural and, in some cases, human waste which they are too miserly to treat properly. Poor countries pollute their marine environments with the wastes that they are too poor to manage in any other way. In some instances the net results are similar and may lead the unwary or the self-interested into the trap of supposing, yet again, that a common effort is needed to cope with a common responsibility. It would be otiose to repeat the now familiar argument against this position and we should look instead at what has already emerged in the way of international agreements and what, if UNEP is taken seriously, we might hope for.[6]

We have already remarked on the EEZs and, so far as they go, they represent a distinct improvement for Third World countries in the international legal framework for the preservation of their own resources. Obviously they can do little to alleviate all the other financial and demographic pressures, but they are something. It is also worth remembering that they were introduced not for this purpose but to solve a problem for the fishing interests of the industrial world. Even among these states agreement is incomplete, as the USA, Germany and the United Kingdom still refuse to ratify the United Nations Convention on the Law of the Sea (UNCLOS). Indeed, at the time of writing only 47 countries had ratified UNCLOS, 13 short of the minimum required, but many of the provisions of that Convention have passed into customary international law. Most of the world's major fishing grounds lie within EEZs, which collectively make up around 45 million square kilometres. Even without the three intransigent nations they may well form one of the bases of sensible international agreements on protecting marine ecosystems simply because they confirm the legal rights of developing as well as developed countries. Just how vital that protection has become may be seen in the statistics for world fish catches (see Table 7.2). These figures, which show how close the actual catch is to the potential limits, make clear that there are not many more fish in the sea.

Table 7.2

Marine fisheries, yield and estimated potential

Ocean/Sea	*Average annual catch (million metric tons)*		
	1977–79	*1987–89*	*Potential*
Atlantic	22.35	21.55	28.23–37.21
Pacific	30.96	48.44	34.11–49.48
Indian Ocean	3.38	5.44	5.31–8.01
Mediterranean/Black Sea	1.21	1.65	1.17–1.53
Antarctic	0.45	0.48	na
Arctic	0.00	0.00	na
World	58.34	77.55	68.82–96.23

Source: extracted from World Resources Institute, *World Resources 1992–93*, Table 23.3, p. 339, Oxford University Press, New York and Oxford 1992.

Altogether there are over 40 international agreements and conventions on the marine environment and on living marine resources. Some are regional, like the Abidjan agreement of 1981 on the protection and development of the coasts of West and Central Africa; some are global, like the International Convention on Civil Liability for Oil Pollution Damage (Brussels, 1969); a few are not in force either because no agreement was finally reached or because the negotiations to persuade enough countries to ratify them are still going on – apart from UNCLOS this last group includes, for example, the Nairobi 1985 Convention covering the East African Regional coasts. A few new instruments are needed, like a settled international convention dealing with the use of mono-filament fishing nets, but by and large if the conventions already in existence were internationally ratified and accepted, then many of the direct legal difficulties would be at an end. More importantly, the way would be paved for future agreements as they become necessary.

At UNCED the Global Forum, in its document on marine pollution, urged NGOs to get behind UNEP and to campaign for the wide acceptance of the regional seas programmes and of conventions like UNCLOS. They also proposed that NGOs should engage heavily in educational and informational programmes and suggested some simple mechanisms for this. Because 'bottom trawling' is so destructive of marine ecosystems, the Forum demanded that it be banned. *Agenda 21*, in a somewhat more elaborate and detailed form, took the same lines in its Chapter 17, and, indeed, set out the need for aid, in one form or another, for the developing world to deal with its polluting problems. It urges the adoption and ratification of a number of existing and crucial agreements and it asked UNEP to 'convene ... an intergovernmental meeting on protection of the marine environment from land-based activities' (paragraph 26), but it failed to specify what it hoped that meeting would do. The agenda listed the causes of world-wide over-fishing:

> management of high seas fisheries ... is inadequate ... and some resources are overutilised. There are problems of unregulated fishing, overcapitalisation, excessive fleet size, vessel reflagging to escape controls, insufficiently selective gear, unreliable databases and lack of sufficient cooperation between States. (paragraph 45)

This, we feel, is a fairly comprehensive condemnation of international industrial fishing practices and, possibly, not a bad subject for the UNEP meeting. Though, for the South the important measures, not dealt with in either set of documents and which UNEP would do well to address, are those which must be taken against poaching. Selling permits to fish in their EEZs to Northern fishing fleets could be a valuable source of income, particularly to small island states, but as long as widespread poaching continues this is a dubious venture.

Because the UN, particularly through UNEP, have got so far despite large financial sectoral interests there is a reasonable hope that many of the marine conventions will eventually be adopted. However, valuable as they may be, they will not deal with the roots of the difficulty. The bulk of marine pollution comes from land-based activities. Northern pollution will continue as long as industry finds it cheaper to pollute than to engage in research into, and the installation of, clean technologies. Southern pollution will continue as long as the developing world lacks the resources to build a clean infrastructure for waste disposal and to adopt cleaner forms of industrial technology. In neither case is destruction positively wilful, though we may make an exception for nuclear bomb tests: in one it is the result of the drive of market forces, in the other it is the result of poverty. The former may be held more or less directly responsible for the latter.

We have already pointed out that the great rivers are the conduits for much of what pollutes coastal waters. This leads us to the other, quantitatively smaller but equally vital wet part of the planet – fresh water. Like the seas, many of the great waterways are shared; 96 of the 170 or so states in the world share just 13 rivers or lakes (see Table 7.3 for the principal examples) and treaties are needed to preserve even the fragile order we at present enjoy. European states, many of whom depend jointly on some rivers, have evolved an immense complex of agreements to support the situation and yet there are still major disagreements over water control, discharges, irrigation and pollution. Few such treaties exist in developing countries and, as the competition for water increases, their lack leaves the way open for serious disputes.

Table 7.3

Rivers and lakes with five or more nations forming part of the basin

Danube	12	Mekong	6
Niger	10	Lake Chad	6
Nile	9	Volta	6
Zaire	9	Ganges-Brahmaputra	5
Rhine	8	Elbe	5
Zambezi	8	La Plata	5
Amazon	7		

Source: P.H. Gleick, 'Climate Change and International Politics: Problems Facing Developing Countries', *Ambio*, vol. 18, no. 6, 1989. Quoted in Robin Clarke, *Water: the International Crisis*, Earthscan Publications Ltd, London 1991.

Even though it can have catastrophic effects on health and well-being, pollution is often not the major issue between riparian states, particularly those of the Third World. Any country which depends for its principal water supplies on rivers which have their origins in states further up-river is, particularly in the absence of treaties, in a vulnerable position. There are plenty of examples of conflict arising out of this situation: water was a significant issue in the war of 1965 between India and Pakistan; Chile and Bolivia have been quarrelling about the River Lauca since 1962; Jordan is in trouble with Syria because of the latter's plans for the River Yarmuk; and Syria and Iraq are both in difficulty because of Turkey's diversion, in 1990, of the Euphrates to fill a reservoir. One of the greatest of all problems is in the threat to Egypt's dependence on the Nile posed by Sudan's needs, which have been increased by a rapidly growing population and by Ethiopia's proposal to take 4 billion cubic metres of the Blue Nile for its own irrigation.[7] There are many more examples throughout the world (see Table 7.4) and they all call for agreements which will ensure, as far as possible, water security for all the parties involved. A glance at Table 7.4 makes it clear that the issues are always very similar, but it is probably too much to hope that the various states will learn from each others' experience.

Globally, fresh water, like food, is not in short supply, it is simply unevenly located but, unlike food, it cannot easily be transported. There have been schemes for towing large icebergs from the poles to some of the drier spots of the globe, but these had too much in common with Icarus' fatal venture. Over three-quarters of all known fresh water lies in glaciers and icecaps and another 22 per cent of it is ground water. The latter is either rain water which has soaked into the underlying rock or what is called 'juvenile' water which arises from deep magmatic sources. Some of this ground water lies 800 metres or more below the earth's surface at the lower depths of the permeable rocks and so is inaccessible, but a substantial amount is reachable by boreholes.

The world's natural water cycle produces about 41,000 cubic kilometres of run-off annually. Just over 7 per cent of this run-off, some 3,000 cubic kilometres, is being consumed; of this about 69 per cent is used for agriculture, 22 per cent for industry and 9 per cent for public or domestic use.[8] By the year 2000 total consumption is expected to rise to as much as 4,500 cubic kilometres, and the proportionate use is expected to change to 55 per cent for agriculture, 29 per cent for industry and 16 per cent for domestic and municipal use. The substantial change projected for domestic use is partly based on the assumption that clean fresh-water supplies will be more commonly available among those at present without them, as well as on the predicted increase in the world's population.[9]

Table 7.4
International water disputes

River	*Countries in dispute*	*Issues*
Nile	Egypt, Ethiopia, Sudan	Siltation, flooding, water flow/diversion
Euphrates, Tigris	Iraq, Syria, Turkey	Reduced water flow, salinisation
Jordan, Yarmuk Litani, West Bank Aquifer	Israel, Jordan, Syria, Lebanon	Water flow diversion
Indus, Sutlei	India, Pakistan	Irrigation
Ganges-Brahmaputra	Bangladesh, India	Siltation, flooding, water flow
Salween	Myanmar, China	Siltation, flooding
Mekong	Cambodia, Laos, Thailand, Viet Nam	Water flow, flooding
Paraná	Argentina, Brazil	Dam, land inundation
Lauca	Bolivia, Chile	Dam, salinisation
Rio Grande, Colorado	Mexico, USA	Salinisation, water flow, agrochemical pollution
Rhine	France, Netherlands, Switzerland, Germany	Industrial pollution
Maas, Schelde	Belgium, Netherlands	Salinisation, industrial pollution
Elbe	Czechoslovakia, Germany	Industrial pollution
Szamos	Hungary, Romania	Industrial pollution

Source: Michael Renner, *National Security: The Economic and Environmental Dimensions*, Worldwatch Paper 89, p. 32, Washington, DC 1989. Quoted in WRI, *World Resources 1992–93*, p. 171, Oxford University Press, New York and Oxford 1992.

Figures like these really only serve to illustrate a difficulty. As we have said, in theory there is enough fresh water in the world, provided by run-off, to support a population some ten times that of the present, but, of course, as we have already observed, it is not evenly distributed. Of the 500,000 cubic kilometres of annual global precipitation only about 20 per cent falls on land. Almost one-third of that falls in the tropical forests of Latin America and the

Caribbean while, for example, less than 1 per cent of it falls in Australia. Within ranges of this kind there are other considerations. If rain falls, as it does in some dry lands, very intensely but very briefly, then the run-off is largely useless. Similarly, some countries which have a reasonably sizeable rainfall may get most of it over a small area – Mexico is a good example of this. However, more and more ground water is being extracted and at a faster rate than is sustainable. By nature it is replenished slowly and, as its surface is the water-table, mining it can cause yet further severe water losses or droughts – rivers and lakes may simply dry up if the water-table is sufficiently lowered. This problem has already been encountered in India's Maharashtra state and around Beijing.

Some of this was partly acknowledged in the UN International Drinking Water Supply and Sanitation Decade (1981–90). That decade was established at a conference held in 1977, in Mar del Plata, where it was resolved that by 1990 everyone should have access to clean drinking-water and proper sanitation. The estimated cost of fulfilling this heroic resolution was put at US$140 billion, of which the UN raised about 25 per cent. Much work was undertaken and by the end of the decade over 700 million people did get safe fresh water supplies and 250 million got effective sanitation. But the under-funding and poor management and maintenance even of what was installed led to widespread failure. At the end of the decade, partly as a result of increased populations, more people needed water and sanitation facilities than at the beginning. By 1990 1.33 billion of the world's population were still without access to safe water and 2.25 billion had no sanitation (see Table 7.5). Bad management and inadequate finance were part of the difficulty, but these were compounded by an imperfect understanding of absolute water shortages and how to cope with them.

In its chapter on fresh water, *Agenda 21* offered a telling, if abbreviated, statistic:

> An estimated 80 per cent of all diseases and over one third of deaths in developing countries are caused by the consumption of contaminated water, and on average as much as one tenth of each person's productive time is sacrificed to water-related diseases. (paragraph 47)

There are some thirty or so waterborne diseases, but the commonest are diarrhoea, typhoid, cholera, yellow fever, roundworm, tapeworm, infections from liver-fluke, guinea-worm, malaria, schistosomiasis (also known as bilharzia) and onchocerciasis (river blindness). In 1992 the World Bank quoted a relatively obvious, but quite useful, review published by the United States Agency for International Development (USAID) which showed the levels of disease reduction achievable by improvements in safe water supplies and sanitation. We reproduce their summary in Table 7.6.[10] If *Agenda 21* is right in its estimate of disease carried by water, then the cuts that many Northern governments are busily making in their aid budgets will help to reduce, among other things, the finance available for necessary safe drinking-water and sanitation projects. This will contribute liberally to the deaths of 4.6 million of the 14 million children who die each year before reaching the age of five.[11]

Table 7.5

Population in developing countries with access to safe water and sanitation

*Profile grouping**	*safe water (%age)*	*sanitation (%age)*
High human development	78	—
Medium human development	78	74
Excluding China	85	—
Low human development	58	33
Excluding India	46	—
All developing countries	68	45
Least developed countries	46	23
Sub-Saharan Africa	40	30

*These are UNDP categories which, although calculated on different bases, roughly correspond to the economic categories used by the World Bank and the World Resources Institute.

Source: extracted from UNDP, *Human Development Report 1992*, Table 2, p. 130, Oxford University Press, New York and Oxford 1992.

Safe fresh water and adequate sanitation, those great Victorian blessings, carry with them the corollary that adequate sewage disposal is essential. Without it not only will coastal pollution continue, but the safety of water supplies will continue to be compromised. During the International Drinking Water Supply and Sanitation Decade many of the sanitation projects failed because they were too expensive, because the scale of what was attempted was too daunting and because they were begun following a long history of under-investment. Most developing countries

Table 7.6
Effects of improved water on sanitation and sickness

Disease	*Millions of people affected by illness*	*Median reduction attributable to improvement (%)*
Diarrhoea	900*	22
Roundworm	900	28
Guinea worm	4	76
Schistosomiasis	200	73

* Refers to number of cases per year

Source: Esrey et al. 1990, quoted in World Bank, *World Development Report 1992*, p. 49, Oxford University Press, New York and Oxford 1992.

are spending only 0.6 per cent of their GDP on sewerage and in Latin America, for instance, only 2 per cent of sewage is treated.[12] Yet there are answers, rather like those in the sustainable agriculture debate. Simple and cheap new technologies have been developed for closed systems which allow for the use or safe disposal of human waste, and there are a number of examples of the cheap construction of sewerage systems of the more familiar kind. An engineer from Recife, José Carlos de Melo, has designed a condominial system for use in residential blocks which cuts normal costs by between 20 and 30 per cent and which now serves several hundred thousand people. There have been failures with this system, but virtually all of them have been due to poor organisation and the failure to involve the people for whom they were being built. In the case of another simple system organised by a community worker in Karachi, Akhter Hameed Khan, cost reduction to below US$50 per household was achieved not just because the system was simple, but by the elimination of corruption.[13]

That the problem was recognised quite early on makes the omission from the Brundtland Report of any discussion of water surprising; it is one of the major factors limiting certain kinds of development in very large parts of the world. Substantial areas of the tropics consist of lands which are arid simply because the climate is dry. Others are subject to periodic droughts, but deforestation, over-grazing and increases in populations relying on limited supplies of water run-off are the principal new factors. Robin Clarke offers a rough guide to determining whether given countries or regions are likely to be facing severe water problems.

He suggests that those 'that use less than 5 per cent of total run-off' are unlikely to be in difficulty; those using between 10 and 20 per cent will 'usually have fairly major water problems'; in any countries using above 20 per cent water supplies will be a constant national problem and may prove to be a substantial limiting factor in development.[14] Available figures obscure differences in the regions of large states like India, but they are nonetheless significant, and in Table 7.7 we may, using Clarke's principle, identify those countries in which water, even at its normally estimated levels when there is no drought, is a major development problem. These have been added to recently by all those Sahelian states now in the grip of major drought, and in some cases of war, which do not normally have a problem.

There are two routes to solutions. Most important is the issue, referred to in Chapter 6, of land use and land rights, of a restoration, in some form, of sustainable farming practices. Central to this route is the need to match the crops to the land in which they are grown; it is frequently the case that inefficient irrigation is used to support inappropriate cash-crops so that one waste of water is added to another. Sugar-cane, for example, is widely grown on Indian irrigated land, but it uses enormous amounts of water. If potatoes were substituted it would be possible, using the same amounts of water, to irrigate between five and ten times the amount of land. Even more could be achieved if these irrigated lands were devoted to growing the Indian staple crops of gram (a group of leguminous plants whose seeds are used as food) and millet. This is a moral, one central to the concept of sustainable agriculture, which is almost universally applicable.

Of increasing international importance is the second of the two routes to sustainability – a mix of water-harvesting and micro-irrigation. There are many ancient techniques for both of these (micro-irrigation means traditional small-scale irrigation as well as the use of plastics and pipes to encourage individual plant growth), many of which have dropped out of use with the advent of the modern pump and must be recovered. Many of the mechanical pumps in use throughout the Third World need expensive maintenance and call for parts which have to be imported from industrialised countries, with a cost in scarce hard currency. But even if this were not a difficulty – after all simpler and locally reparable pumps can be built – in many instances ground-water supplies are being used faster than they are replenished and problems multiply as a consequence.

Table 7.7

Freshwater resources and withdrawals in water-stressed countries, 1990

Country	*Total renewable annual resources (cubic km)*	*Annual withdrawals (cubic km)*	*%age of resources*
Africa (excluding South Africa)			
Algeria	19.10	3.00	16
Cape Verde	0.20	0.04	20
Egypt	58.30	56.40	97
Libya	0.70	2.83	404
Madagascar	40.00	16.30	41
Mauritania	7.40	0.73	10
Mauritius	2.20	0.36	16
Morocco	30.00	11.00	37
Sudan	130.00	18.60	14
Tunisia	4.35	2.30	53
North & Central America (excluding US)			
Barbados	0.05	0.03	51
Cuba	34.50	8.10	23
Dominican Rep.	20.00	2.97	15
Mexico	357.40	54.20	15
South America			
Peru	40.00	6.10	15
Asia (excluding Japan, Qatar, Saudi Arabia and the United Arab Emirates)			
Afghanistan	50.00	26.11	52
China	2,800.00	460.00	16
Cyprus	0.90	0.54	60
India	2,085.00	380.00	18
Iran	117.50	45.40	39
Iraq	100.00	42.80	43
Israel	2.15	1.90	88
Jordan	1.10	0.45	41
Korea (N.)	67.00	14.16	21
Korea (S.)	63.00	10.70	17
Lebanon	4.80	0.75	16
Oman	2.00	0.48	24
Pakistan	468.00	153.40	33
Thailand	179.00	31.90	18
Yemen			
(Arab Rep.)	1.00	1.47	147
(People's Dem.)	1.50	1.93	129

Source: extracted from WRI, *World Resources Report 1992–93*, Table 22.1, pp. 328–9, Oxford University Press, New York and Oxford 1992.

Water-harvesting has been practised for as long as people have inhabited dry lands and we sometimes fail to appreciate the extent of it. The Negev desert, for instance, is not significantly drier now than it was before the Yarkon river irrigation scheme was initiated[15] and yet it supported thousands of small farms of between 0.5 and 2 hectares. Crops which grow quickly and the careful use of water catchments were then and still are the answer. Other practices, like ensuring the growth of live hedges in certain places, building bunds for water catchment, as in the famous example at Yatenga in Burkina Faso, mulching and dew-trapping can increase the productivity of all but the most desertified of dry lands.

Not all the answers can be provided simply by some return to ancient and sustainable practices. It is clear that in spite of their use many communities perished in the past, often for political reasons but also because of natural climate stresses and because of poor management. Hence the importance of adding a few modern techniques to traditional ways which can produce some startling results. During the 1970s when it became clear in the Sahel that previously normal rainfalls would not recur, Niger's food production had fallen by at least 30 per cent and very many of its people were destitute. Lutheran World Relief (LWF), a major NGO with an impressive record of working collectively with some of the world's poorest people, set out to help local communities to improve their rates of food production. Some of Niger's desperately poor were nomads unable to continue their ancient way of life any longer because of the Sahelian drought, others were settled small farmers. Both groups were to be found struggling for survival, for the usual reasons, on poor, marginal land. LWF's programme was designed not only to satisfy their immediate basic needs but also to attempt the regeneration of their environment.

Before LWF began its work some villagers had built shallow wells which did produce a certain amount of polluted water, but which also tended to collapse. Advancing sand-dunes and hot, eroding winds made the growing of vegetables particularly difficult, but in any case the villagers lacked seeds for more than a very few species and, because they had lost their livestock in the drought, they had virtually no natural fertiliser. The first steps consisted of supplying a greater variety of seeds and in assisting with the introduction of chemical fertilisers. Impressive amounts of food were produced but they were accompanied by new problems. Insect pests attacked the plants and the soil immediately became infested with nematodes (a distinct group of round-, thread- and eel-worms), a threat both to plants and to health. Limited quantities of pesticides and careful instruction in their use were then provided, together with encouragement to destroy as many pests as possible by hand.

Planting legumes, like chick-peas, a pre-drought practice, was reintroduced by the villagers and these were not only an easily dried, stored and replantable source of food but, when intercropped with other vegetables, fixed nitrogen in the soil which helped to destroy the nematodes. Ponds were recovered and traditional varieties of trees and shrubs were planted around them, thus reducing the rate of evaporation and preventing depredations by the remaining livestock. All this not only allowed the enormously extended cultivation of green vegetables, but the sustainable use of water considerably reduced consumption.

The final phase of the programme was to help in solving the difficulties met by the villagers in building shallow wells. For this steel-reinforced concrete rings, 1 metre long and 1.4 metres in diameter, were developed. These enabled the building of simple culverts to reach water-tables often at no greater depth than about 6 metres, thus giving villagers access to clean water. In 1986 the cost of the wells, including labour, was about US$400, substantially less than those provided by the government of Niger, and they have a life expectancy of about 50 years. By 1987 3,200 such wells had been constructed and the programme had successfully demonstrated one answer to the problems of drought.[16] Obviously these techniques are valuable in regenerating degraded land and in making the most effective use of scarce water whether or not there is a drought. So far as rural areas are concerned the revival of older sustainable techniques together with a few technologically simple modern improvements could go a long way towards dealing with a very ancient difficulty. But it is not possible to consider safe drinking-water, water for agriculture and issues to do with health and sanitation in isolation from the central questions raised in Chapters 5 and 6 .

After agriculture the next largest demand on the world's water supplies comes from industry. Past practice throughout the world has led to the construction of industrial plants wherever there is sufficient water for them; worrying about the problems then created comes later:

> In a sample of fish and shellfish caught in Jakarta Bay, Indonesia, 44 per cent exceeded WHO guidelines for lead, 38 per cent those for mercury and 76 per cent those for cadmium ... During the 1980s lead also worsened or became a problem for the first time in some rivers in Brazil (Paraíba and Guandu), Korea (Han), and Turkey (Sakarya).[17]

Pollution is one difficulty which other users of industrial water sources have had to face, but for much of the world the shortage of water is an even more serious issue. In the construction of developmental projects fuel, transport and the availability of labour have usually been considered, and it is now becoming increasingly important to take water supplies into account. Above all more attention must be given to recycling industrial water either for further industrial use or for return to the water-courses. Major difficulties exist in this area because of sectoral cost accounting. Robin Clarke quotes the case of an estimate, made in California in 1980, that the value added to every cubic kilometre of water used in spraying crops was US$75 million, but every cubic kilometre used by industry resulted in an added value of $65 billion. He goes on to say that the Chinese have also calculated that water used in industry is, in simple economic terms, 60 times more valuable than when it is used in agriculture.[18] Countries faced with poverty and debt will increasingly be tempted into development projects built as cheaply as possible and making unsustainable demands on their water supplies.

Perhaps in order to make up for Brundtland's lack, the chapter on fresh water in *Agenda 21* is one of its longest. It begins with a fairly sensible demand, made on all states, that proper assessments of needs and supplies should be made, that sensible water management schemes should be instituted and that research and information should be widespread and exchanged. It appeals for the 'Development of new and alternative sources of water-supply such as sea-water desalination, artificial groundwater recharge, use of marginal-quality water, waste-water reuse and water recycling' (paragraph 12 [j]). Magnificently missing the point, it also calls for the delegation of water management to the 'lowest appropriate level' (paragraph 12 [o.i]). We must assume that the document is the work of many hands; not all references to the communities in need of water are quite so crass and in paragraph 65, in the context of water needs in sustainable agriculture, we have something like a manifesto:

> The challenge is to develop and apply water-saving technology and management methods and, through capacity-building, enable communities to introduce institutions and incentives for the rural population to adopt new approaches ... The rural population must also have better access to a potable water-supply and to sanitation services. It is an immense task but not an impossible one ...

So long as everyone recognises that 'communities' includes local communities and that populations, urban as well as rural, must be enabled to play their part in planning and managing water resources, then our task is to persuade those governments which endorsed *Agenda 21* at UNCED actually to bring its programme into being. 'An immense task but [we fondly hope] not an impossible one.'

8
Waste

We are choking, frying and drowning in our own waste. There are austere souls who, lacking a religion to do the job for them, spend much time in trying to find a way of distinguishing humanity from animals. One, which they have tended to overlook, lies in the phenomenon of waste – we seem to be the only creatures able so to transform and concentrate matter that it cannot be reabsorbed into any other chain or ecosystem without tending to degrade or destroy it. In the North we engage in this transformation on a mammoth scale (see Table 8.1 for the quantities of municipal wastes alone) and then expend much time, money and ingenuity in storing those parts of it that we have not managed surreptitiously to dump in a hole in the ground, in the nearest sea or on some unsuspecting Third World country. We also burn quite a lot of it and tend to be outraged when those living near the fire complain of death, disease and deformity – they were obviously irresponsible in choosing to live in such an unhealthy area.

In its *World Development Report 1992*, the World Bank offers some telling comparisons (p. 54). It begins by remarking that the production of 'hazardous materials and wastes' is on the increase, but yet again underlines the point that it is within the industries of the developed world that the problem begins:

> Industrial economies typically produce about 5,000 tons for every billion dollars of GDP, while for many developing countries the total amount may only be a few hundred tons. Singapore and Hong Kong combined generate more toxic heavy metals ... than all of Sub-Saharan Africa (excluding South Africa).

Table 8.1
Municipal waste in OECD countries

Country and years of estimate	*Annual municipal waste generation (year of estimate in bold)*		*Composition (%age of total weight) (year of estimate in italic)*					
	(000 metric tons)	*per capita (kg)*	*paper and cardb'd*	*plastic*	*glass*	*metal*	*organic as %age of other*	*inorganic*
Australia **1980** *1980*	10,000	681	26.0	6.1	15.1	7.0	45.8	41.4
Austria **1988** *1985*	2,700	355	33.6	7.0	10.4	3.7	45.3	60.5
Belgium **1989** *1989*	3,470	349	28.3	7.7	7.6	3.7	52.7	47.6
Canada **1989** *1989*	16,000	625	36.5	4.7	6.6	6.6	45.6	74.3
Denmark **1985** *1985*	2,400	469	38.6	3.4	5.4	5.0	47.6	81.3
Finland **1989** *1985*	2,500	504	40.0	8.0	4.0	3.0	45.0	85.0
France **1989** *1989*	17,000	303	27.5	4.5	7.5	6.5	54.0	59.0
W. Germany **1987** *1985*	19,483	318	17.9	5.4	9.2	3.2	64.3	63.4
Greece **1989** *1989*	3,147	259	20.0	7.0	3.0	4.0	66.0	57.0
Iceland **1985** *na*	93	386	na	na	na	na	na	na
Ireland **1984** *1984*	1,100	311	24.5	14.0	7.5	3.0	51.0	56.0
Italy **1989** *1986*	17,300	301	22.3	7.2	6.2	3.1	61.2	64.4
Japan **1988** *1989*	48,283	394	45.5	8.3	1.0	1.3	43.9	77.2
Luxembourg **1990** *1985*	170	466	17.2	6.4	7.2	2.6	66.6	44.0
Netherlands **1988** *1988*	6,900	465	24.2	7.1	7.2	3.2	58.3	88.3
New Zealand **1982** *1980*	2,106	670	33.6	3.0	2.5	7.6	53.3	37.0
Norway **1989** *1988*	2,000	473	30.0	5.0	3.0	7.0	55.0	77.0

Table 8.1 (continued)

Country and years of estimate	*Annual municipal waste generation (year of estimate in bold)*		*Composition (%age of total weight) (year of estimate in italic)*					
	(000 metric tons)	*per capita (kg)*	*paper and cardb'd*	*plastic*	*glass*	*metal*	*organic*	*inorganic as %age of other*
Portugal **1985** *1985*	2,350	231	19.0	3.0	3.0	3.5	71.5	74.5
Spain **1988** *1989*	12,546	322	20.0	7.0	6.0	4.0	63.0	49.0
Sweden **1985** *1980*	2,650	317	43.0	10.0	5.0	6.0	36.0	89.0
Switzerland **1989** *1989*	2,850	424	32.0	13.0	7.0	6.0	42.0	70.0
Turkey **1989** *na*	19,500	353	na	na	na	na	na	na
UK (exc . Scotland) **1989** *1980*	18,000	357	29.0	7.0	10.0	8.0	46.0	58.0
USA **1986** *1984*	208,760	864	34.7	6.7	9.0	8.8	40.8	37.5

Source: extracted from World Resources Institute, *World Resources 1992–93*, Table 21.4, p. 319, Oxford University Press, New York and Oxford 1992.

This section of the report goes on to say that although the production of toxic wastes in the developing world is not yet an international problem of any great importance, with increased industrialisation it could become so. As an example it offers Thailand, which in 1969 boasted only some 500 factories but now (1993) has over 26,000 all producing hazardous wastes. It also predicts that, if present trends continue, 'the volume of toxic heavy metals generated in countries as diverse as China, India, Korea, and Turkey will reach levels comparable with those of present-day France and the United Kingdom within fifteen years' (pp. 54–5).

Waste pollutes and it can do so on a vast scale. We considered some aspects of the problem in that part of Chapter 7 which dealt with the oceans and seas. It is enough to say here that there are virtually no unpolluted coasts anywhere in the world. Some forms of waste are obvious enough, we can see them in everything from the discarded champagne corks in the Ascot Royal Enclosure to the plastic wrappings lying in the street around the local hamburger shop or, more horribly, the stacking of the pit spoil which crushed or drowned hundreds in the Welsh mining village of Aberfan in 1966. Others are more insidious, like the airborne chemical emissions which go towards creating acid precipitation or the nuclear waste from Sellafield which makes fish from the Irish Sea such a doubtful source of nutrition. Waste also pollutes not just because individual items are discarded, but because they are frequently and carelessly dumped in combinations which produce new hazards.

Untreated or partly treated sewage discharged into rivers and seas is one of the greatest forms of polluting waste and it leads not only to the spread of disease, the disturbance and even destruction of marine ecosystems, but also to the waste of water, an increasingly scarce resource, and the waste of usable biological material. Technologies exist for dealing with this problem, as with so many others, in environmentally sustainable ways, but their use on a scale wide enough throughout the world to make a difference would be, in the present forms of accounting, very expensive. Industrial waste, including packaging, is made up either of materials which are cheaper to throw away than to recycle or materials which, at least at present, no-one really knows what to do with. The most obvious example of the latter is nuclear waste, a term which covers an enormous range of materials.

Quite recently a palliative for the problem was proposed by a group of wide-eyed and enthusiastic economists from an organisation known as the London Environmental Economics Centre (LEEC). They pointed out that pollution is caused not just by what might normally be called waste, but also as a consequence of creating some other kind of good which incidentally has the effect of damaging the environment. Any industrial development, no matter how carefully controlled, will have some ill effect on the environment, and the assumption is usually that the good resulting from these things will outweigh the damage. The issues, as our economists remarked, are clear; who benefits and who suffers, how great is the damage and by what means do we measure it, are all questions in need of an answer.[1] Very roughly, the same questions surround the more or less careful disposal of waste.

The principal measure for calculating the answers to these questions is money. If the estimated financial gain from any given project is sufficiently greater than its total cost, then, in general, it will go ahead. As clean air, unpolluted fresh water, a beautiful view or even the graves of our ancestors do not, as a rule, have a market price it is not easy to put them in the balance. The people from LEEC, bringing together a growing consensus among economists, proposed ways in which this lack could be met and by which things normally thought of as freely available could be given a monetary value. This, they claim, would serve two purposes: it would allow the true environmental costs to be assessed, which might redress the balance, and it would show what the developers should pay for the right to go ahead.

The LEEC team extended the principle. Some continued pollution, particularly by waste products from energy production and industry, is always to be expected, but it must be made containable. To get to the point where it is cheaper to install the technology to deal with it than not, a tax, equivalent to the cost of cleaning up the mess, should be levied. The idea has become very fashionable and politicians regularly talk grandly of 'carbon taxes', even if they show few signs of introducing them. It is possible, of course, that they have been frightened off by the TNCs: Shell, Hoogovens (steel producers), the KNP paper mill, Akzo, DSM and Dow (all chemical companies) and Hoechst have all threatened the Netherlands with factory closures if carbon taxes are introduced.[2]

Despite TNC pressures some countries have begun to introduce cost–benefit analyses which might provide the basis for such taxes (among them, for example, are Germany, Italy, the Netherlands, Norway and the USA) and industry may yet begin to pay up, in some proportion at least, for the damage it does with its wastes. It is all an excellent scheme and really only suffers from two flaws. The first is the universal government habit of regarding all particular taxes as part of general taxation and not applying them to the designated purpose; the second is that, in general, industry will find it easier to pass the cost of the taxes on to its consumers than to invest in clean production. As our economists would be the first to point out, these are political rather than economic objections and so outside their province. George Bernard Shaw was credited with the remark: 'If all economists were laid end to end, they would not reach a conclusion.'

We should not cavil; perhaps any advance is better than none and at the very least LEEC reminded the politicians that waste and pollution are real problems that will not simply go away. There is a tendency to divide waste into two major classes: toxic and non-toxic. While it is unquestionably true that there are some wastes which are dangerous or toxic on their own, virtually any waste can cause some damage and, even more seriously, can interact with others to become toxic. One current estimate of the levels of toxic waste production, offered to the Brundtland Commission, suggested 375 million tonnes per annum,[3] but this is a very unreliable figure since another analysis, confined to nineteen countries only, suggested that between them they were producing nearly 500 million tonnes.[4] There are equally dubious estimates for the world-wide production of 'non-toxic' wastes. Each year OECD states produce 1,430 million tonnes of industrial waste each year (as well as releasing into the atmosphere 76.5 million tonnes of chemicals which add to acid rain and 2.3 billion tonnes of carbon dioxide and other greenhouse gases).[5] The figures for the OECD municipal waste given in Table 8.1 do not distinguish between 'toxic' and 'non-toxic' materials, but they total over 420 million tonnes per annum. All these figures provide us with a rough guide to the world's waste problem, and no matter how unreliable they may be they do illustrate the mammoth scale on which waste is generated. Some further measure of it may be seen in the profits derived from its disposal. For example, in 1987 the largest US waste haulage company, Waste Management, made US$2,757 million profit. Attwoods, once a small British company in the same business and now a substantial operator in the USA (in the same year it made $168 million), imports refuse from the United States into Britain, where waste is also big business; Atwoods has estimated that the total expenditure on all forms of waste control in Britain is around $5 billion a year.

We tend to think of toxic waste, when we think of it at all, as something rather special, but it is often fairly ordinary, even domestic. For example, in the United States alone some 2.5 billion exhausted batteries are thrown out annually and each tonne of them contains, among other things, 270 kg of manganese dioxide, 210 kg of iron, 160 kg of zinc, 20 kg of copper and nearly 2 kg of mercury. These are all heavy metals and if those batteries are not carefully dealt with they can easily become environmental hazards. If, as is commonly the case, they are dumped in a hole in the ground – what is usually called a 'land-fill site' – they can leach into ground water and may well contaminate supplies of drinking water.

Mercury, in particular, is dangerous and one of the nastiest cases of poisoning produced by it, though not caused by the careless dumping of batteries, happened in the Japanese town of Minamata when, between the years of 1953 and 1956, forty-three people were killed and many more were severely and permanently injured or suffered from brain-damage. Women gave birth to deformed babies and there was a considerable increase in genetic malformations. Fish caught locally had been contaminated by substantial discharges of dimethyl mercury from a nearby plastics factory and those who ate them ingested the poison. As mercury cannot be excreted, large amounts of it accumulated in the bodies of those who ate the fish, with these dramatic and tragic results.[6]

The number of discarded products which are toxic by themselves is legion, and we need here only consider one other example. We have already remarked that nuclear waste covers a huge range of items, including everything from spent fuel to contaminated overalls, from tailings from uranium mines to certain used medical equipment from hospitals. A distinction is usually made between high- and low-level wastes, though this can obscure some major dangers. Projections based on current production suggest that by the year 2000, the radioactivity of world-wide stocks of high-level waste will reach 150 billion curies.[7] The substances vary and radiate energy from their nuclei at differing rates – iodine-131, for example, has a half-life of only 8.6 days, while plutonium has a half-life of 24,400 years and uranium-238 a half-life of 4.5 billion years.[8] These substances are made up of gases, solids and liquids, each presenting their own problems of disposal. For the longer lived agents no serious meaning can be given to the idea of 'permanent' disposal.

Very little waste radioactive gas is produced and it is not generally stored. It is normally released into the atmosphere in quantities small enough not to be a danger. That information, of course, comes from the usual and totally reliable sources within the industry that gave us Sellafield, Chernobyl and Three Mile Island. So far as the present authors are aware, no proper scientific assessment of these risks has ever been made.

Most solid wastes in need of disposal come from nuclear power stations (which at present produce a little more than 2 per cent of the world's energy)[9] and are made up either of irradiated fuel elements or of scrapped equipment. Some of this material can be incinerated at very high temperatures, in the process allowing the recovery of valuable, and exceedingly dangerous, plutonium. Much of it, however, must be buried, usually in underground containers but sometimes in weighted containers at sea. Both of these forms of disposal have aroused considerable concern because there is no conceivable way in which their long-term safety can be guaranteed; their use simply passes the problem on to future generations. Burial at sea is even more controversial as, at least in theory, the seas are common while the nuclear rubbish is not. For any nuclear nation to endanger this 'common' is clearly immoral and ought to be illegal.

Even greater difficulties arise in the disposal of liquid nuclear wastes. Liquids can leak and no-one has yet designed a container guaranteed to last in good condition for the 10,000 years or so that it will take before the waste is safe. We may even take leave to doubt the recent claims of those who feel that they could design one to last safely for 1,000 years. A controlled experiment to determine the accuracy of the claim might prove difficult to administer; after all even Emilia Marty, Janáček's heroine in *The Makropoulos Case,* only lived for 300 years. There is some hope that the liquids might be vitrified, packed into titanium-cobalt containers and buried like solid waste. Dumping these packages on to other planets has also been suggested, but, if for no other reason, the failure rate of space-shuttles and rockets makes this solution a little shaky.

These are just examples which are not necessarily representative because the toxic waste products of industry are almost as varied as industry itself. Many of them cannot even accurately be described since they are made up of new and unstable chemical compounds whose properties are imperfectly understood. We have remarked before that even non-toxic waste can be dangerous if it is not properly disposed of – we have only to think of the health-threatening properties of ordinary kitchen waste if it is simply left to rot. It is often very difficult to dispose of dangerous material safely; we have pointed to nuclear waste, which must be the most grotesque example, but there are many others. Among disposal techniques incineration, for instance, no matter how carefully managed, regularly produces dioxins and furans which are frequently discharged into the atmosphere and are responsible for a host of life-threatening illnesses and for birth defects. Claims that modern incinerators do not pollute come from the same kind of worthy spokespeople as the public relations employees of the nuclear industry. We are in the same boat as Matilda's aunt:

Her Aunt, who, from her Earliest Youth,
Had kept a Strict Regard for Truth,
Attempted to believe Matilda:
The effort very nearly killed her.[10]

Reckless credulity could be the death of us.

Widespread mining, quarrying, abandoned industrial activity of one kind or another, has left the developed world with a substantial number of holes in the ground. To these may be added a certain number of natural fissures that no one really cares about. Much waste of all kinds has been dumped in these, not always with much regard to safety. This has often meant serious problems caused by poisonous substances leaching from these dumps into ground water supplies, or from the build-up of dangerous gases produced by chemical reactions between the various wastes thus lumped together. Many of them are also fire hazards. This method of disposal is dangerous and uncertain, but in the North it is also rapidly becoming impossible as many countries run out of suitable holes.

The problem is global and has to be tackled at two levels. First, more technologies which are simply less wasteful must be developed: in Chapter 2 we pointed to reasons why one of the best starting points for such development would be in energy production. Second, waste is just that; in an enormous number of cases its components can, with care, be recycled or, with the aid of new technologies, can be substantially reduced in volume (see Table 8.2 for a recent estimate of such possibilities). At the very least much of it can be burnt to produce energy, an option which is slowly gaining ground. Figures produced in 1985 showed that Sweden, Denmark and Luxembourg now recycle between 50 and 76 per cent of their waste to create energy. The Netherlands, France and what was West Germany manage to use between 18 and 27 per cent, while Britain manages a mere 6 per cent and Italy only 4 per cent.[11] Once again the reasons for doing far more to recycle for all purposes are not simply to do with reducing the problem of waste to manageable proportions, but also to do with the preservation and sustainable use of natural resources of all kinds. In many cases recycling technologies are well known, but industries will frequently not make use of them simply because they demand dividend-threatening investment.

Table 8.2
Potential for waste reduction through low-level practices, Germany

Type of waste	*Amount of waste, 1983 (milllions of tons)*	*Potential waste reduction (per cent)*
Sulphurous (acids, gypsum)	2.2	80
Emulsion	0.5	40–50
Dyes and paint residues	0.3	60–70
Solvents	0.3	60–70
Salt slags	0.2	100
Other wastes	1.2	low
Total	4.9	50–60

Source: OECD, *The State of the Environment*, Annual Report, Paris 1991, quoted in World Bank, *World Development Report 1992*, p. 129, Oxford University Press, New York and Oxford 1992.

However, there are some encouraging straws in the wind, particularly in the USA. Two TNCs, the Minnesota Mining Company (3M) and the Monsanto Company, have made dramatic promises to reduce their wasteful and polluting ways. One, 3M, has introduced new technologies to its production systems and has discovered not merely that it can reduce wastes, but that doing so is cheaper. Monsanto undertook to reduce its polluting emissions to zero by the end of 1992. (At the time of going to press in May 1993, we were unable to discover whether this target had been achieved.) Cost-saving is one obvious incentive to industrial waste production, but the USA is alone in having another. As a consequence of environmental lobbying, Congress has begun to enact legislation demanding full disclosure of waste-producing and polluting activities. Many companies are uneasy at the prospect of the publicity that their more unattractive ways might bring.[12]

In the meantime, in the developed world, even in those countries like the United Kingdom which define democracy somewhat more narrowly than others (the UK ranks sixteenth in UNDP's Human Freedom Index, barely ahead of Greece and Costa Rica),[13] popular protest has at least slowed down the random dumping of waste and ensured that some financial and research effort is put into rendering it safe. Yet shareholders' interests have to be protected and waste must be disposed of as cheaply as possible. Fortunately industry has not been slow to protect these interests and two, related, strategies have emerged to ensure that the dividends keep rolling.

One of these is, wherever possible, to move production to the Third World, where governments have not been able, or simply have not known enough to be able, or have been too careless of the poor, to construct legislation to protect workers and the general public from the polluting and otherwise dangerous activities of these unregulated corporations. A notorious example of this is to be seen in Brazil in the city of Cubatao, which has long been known as 'The Valley of Death'. It lies quite close to the major port of Santos and among its industries are Brazil's largest petro-chemical plant, the French Rhodia company and a pesticide factory owned by Union Carbide. In 1981 an estimated 473 tonnes of carbon monoxide, 182 tonnes of sulphur, 148 tonnes of dust and 41 tonnes of nitrogen oxide were emitted in Cubatao. In February 1984 a pipeline carrying gasoline leaked into a nearby swamp over which a shanty town constructed on stilts had been built. The swamp caught fire and at least 500 of the inhabitants were killed and many more seriously injured. In January 1985 a pipeline from a pesticide plant burst, releasing fifteen tonnes of liquid ammonia. Fortunately no one died but 5,000 people were evacuated and 400 of them had to be given oxygen at first aid posts. By the early 1980s 44 per cent of the residents of Cubatao suffered from the respiratory diseases of tuberculosis, pneumonia, bronchitis, emphysema or asthma. Twelve out of every 10,000 babies were born without brains and 8 per cent of the children had spinal or other bone deformities. People living in Cubatao have a life expectancy of 30 years, just under half the national average of 65.6 years.[14] Formerly living and productive rivers have been poisoned and much of the vegetation in the area has died. These levels of unregulated and polluting waste have grown unchecked since the city was developed in 1964. Finally, in 1985, as a consequence of local pressure, the state government filed suits against some of the largest polluters.[15] Other victims of industrial pollution are less fortunate.

The second major strategy developed by Northern industrial countries for dealing with their waste problems was, even according to standards of merchant barons, a touch on the rough side. It consisted of striking deals either in or with Third World states where import controls are inadequate, or where knowledge of dangerous substances is lacking. Either the state or some individual entrepreneur was paid to accept dangerous and usually badly packaged materials for storage or even for 'reprocessing'. The deals often involved several intermediate companies, some existing only on paper, and shipments were frequently toxic cocktails of mixed and poorly described materials crossing several borders on their way to their destination. Perhaps the most notorious example of this charming little device was the Koko dump and *Karin B* story. Despite its fame it is worth reminding ourselves of the salient points.

In 1987 an Italian businessman sought customs clearance in Nigeria permitting him to import 'non-explosive, non-radioactive and non-self-combusting' materials for use by his construction company. They were exported from Italy but originated from the United Kingdom, the USA, the Netherlands, France and Norway as well as from Italy. These 'construction' materials were, in fact, chemical waste containing highly inflammable solvents and PCBs and they arrived in Nigeria in five shiploads during 1987–8. Forged customs documents gave the cargoes, which included 400 drums labelled 'frozen orange juice', clean bills of health. Our Italian businessperson had, for about US$100 per month, rented a plot of land in Koko from a local farmer on which to store some 8,000 barrels holding 3,000 tonnes of dangerous chemical waste. One fairly reliable and subsequent estimate has put the cost of disposing of those wastes in their countries of origin at about $1,750 per tonne, a total of $5.25 million. The Italian's company, after defraying transport and 'incidental' expenses, made a profit on the deal of over $600,000.

Nigerian students uncovered the scandal and pressed the government into action. Three independent teams, one hired by Friends of the Earth, one by the UK Atomic Energy Authority and one from the United States Environment Protection Agency, examined the waste and identified its dangerous contents, which had begun to leach both into the sea and into the local aquifers. Nigeria then tried to persuade the Italian authorities to take back these wastes which had, after all, been dumped by one of their nationals. It was not until the Nigerian authorities took as hostage an uninvolved Italian ship and suspended diplomatic relations with Italy that the Italian government agreed to take some responsibility. Two ships, one called the *Deep Sea Carrier* and the other the *Karin B* were chartered to pick up the waste. The *Karin B* carried 2,100 tonnes of it.

The *Karin B* then sailed enormous distances from port to port looking for a country which would agree to accept its cargo. It failed to do so and finally returned to Italy. Here the limits to Italy's acceptance of responsibility were discovered: since the Italians did not have the facilities to deal with the wastes concerned, they could not be accepted for importation. Britain, that ultimate home of lost garbage, considered taking the cargo, claiming to have the capacity to deal with it. Meanwhile the 150 Nigerian workers employed to clean up Koko suffered from burns, nausea, vomiting and partial paralysis, and some mothers in Koko gave birth prematurely.

This case is worth recalling yet again simply because it involved so much crooked dealing. Not only were the customs documents forged, but at least one of the Italian companies concerned gave a false address and may even have not existed. Nonetheless this shady set-up was shifting wastes for perfectly 'respectable' Northern corporations. Other industrialists in the developed world with dangerous waste to dump simply rely on not being found out. For example, in July 1988 people living near a dump by the River Freetown in Sierra Leone began to complain that their eyes were watering and that they were choking on fumes from the dump. Police investigations led to the discovery of 625 bags of toxic waste that had been dumped there by a British company.

There are other twists to this kind of exploitation. Fly-ash, the irreducible product of incinerators, is not classified as 'toxic' or 'dangerous' under the Environmental Protection Act of the USA and so there is no bar to exporting it. Not everyone takes quite such an insouciant position on the matter. Efficiently run incinerators should, in theory, be able to reduce what they burn to between 10 and 30 per cent of its original volume. However, municipal solid waste is a mix of items of different sizes, densities and materials. Some of them are inflammable, like plastics; others, like glass, are largely incombustible. Because municipal wastes include food residues as well as, for example, discarded metal objects, the water content will vary. No one machine can deal effectively with such a mix and almost every sample of incinerator ash so far examined has traces of polychlorinated dibenzo-p-dioxins (PCDDs – dioxins) and polychlorinated dibenzofurans (PCDFs – furans). Even crematoria which burn only bodies and their coffins, largely organic matter, have been known to produce dioxin residues.

So when, on a small island off Guinea, the trees began to shrivel and investigations led to 15,000 tonnes of incinerator ash which had been sent by the USA on a Norwegian ship called the *Bark*, alarm bells began to ring. The ash had been picked up in Philadelphia and taken to Panama, where the authorities, warned by Greenpeace, refused to allow it to be discharged. Ownership of the ash was then transferred to Alco Guinée, a firm based in Guinea; it was described in the *Bark*'s manifest documents as 'raw materials for bricks' and the import was permitted.

Following the local complaints the government of Guinea began to investigate and a survey of the site showed that the ash had been unloaded in a quarry which at one point came within 10 metres of the sea. There were two cracks in the walls of the quarry through which, particularly during heavy rains, the ash could be washed out to sea and there was a strong possibility of it also seeping into the local aquifers. As a result of information about the shipment laid by Greenpeace before Guinea's ambassador to the USA, the government finally demanded that the Norwegian shipowners should remove the ash. They refused, claiming that it was no longer their property and that it belonged to Alco Guinée, whose responsibility it must be. The government responded by arresting Sigmund Stromme, who was the Norwegian honorary consul and also the company's agent. Several officials of the Guinea Ministry of Commerce were also arrested and four of them were given four-year prison sentences. Stromme was fined US$600 and given a suspended sentence of six-months' imprisonment. This persuaded the Norwegian government to bring pressure on the carrying company which then chartered another ship, the *Banya*, to pick up the ash and take it back to the USA. There the cargo, not considered toxic when it was exported, was refused entry for some time while the Environmental Protection Agency decided whether or not it was dangerous.

Northern worries about the problems of waste are growing rapidly and the EC has begun to introduce regulations governing the movement of all sorts of hazardous substances, but particularly of waste, within the Community. It would be comforting to think that this concern extended beyond Northern borders, but recently the press has reported worrying rumours, emanating from the World Bank, that developing countries are to be encouraged to earn hard currency by making sites available for waste-dumping. The OECD has estimated that Europe alone annually exports about 20 million tonnes of hazardous waste to the developing world. In a shifting universe it is, we suppose, cheering that some Northern attitudes remain constant. Nonetheless, concern is growing. Waste is increasingly recognised as a hazard to health and safety, particularly in the atmosphere and in rivers and seas, but also, now, in all its land-based forms. It is also becoming widely accepted that exporting the problem to the poor is not an acceptable answer and that it simply adds to some of the difficulties that UNCED was supposed to address.

Closely allied to the question of toxic waste is that of dangerous chemical compounds and their movement. The chemical industry is vast and curiously reticent about much of its business. Part of that secrecy is due to the intense competition between corporations, but part of it is also due to the very dangerous nature of much of the industry's produce. Instances of chemical-industrial accidents are legion and, of course, Bhopal is the most infamous of them all, but we should also call to mind just a few others as illustration of their continued frequency. Among these, in the 1970s and 1980s, was the explosion in July 1976 in the Hoffman-La Roche Givaudan chemical plant in Seveso and its appalling and continuing consequences; in November 1979, in Mississauga, Ontario, twenty-one railway goods wagons containing caustic soda, chlorine, propane, styrene and toluene were derailed and exploded, and a quarter of a million people had to be evacuated; another derailment, this time of two wagons carrying chlorine, took place in Montana, Mexico, in August 1981, when 90 tonnes of the chemical were released and 29 people were killed, over 1,000 injured and some 5,000 evacuated; in January 1985, shortly after the explosion at Bhopal, there was a leak of toxic chlorine fumes from a dyeing works at Koratty in Kerala; in 1986 a fire at a chemical warehouse in Basel, owned by Sandoz Inc., caused the release into the Rhine of massive amounts of toxic chemicals including 30,000 kg of pesticides – at about the same time Ciba-Geigy simply dumped a smaller quantity of pesticides in the river.[16]

These accidents are only random examples. The US Environmental Protection Agency compiled a list of chemical accidents which occurred in the USA between the years of 1980 and 1985. They listed 6,928 accidents in the period, in which 139 people died, 1,478 were injured, 217,457 were evacuated and over 190,000 tonnes of chemicals were released. It is said by its compilers that the list only reflects reports from sources immediately available to them and that the real totals are all substantially higher.[17] For the rest of the world no serious figures exist.

For many companies the cost of accidents is outweighed by the cost of safety. Not only is it cheaper to compensate but, given legal uncertainties, compensation may even not be necessary or may be delayed by so many years that interest on the capital that would otherwise go towards it, together with inflation, can take care of the costs. The views of the industry may be seen in remarks, made in Ireland in 1980, by the managing director of Gaeleo in Cork, a subsidiary of the Swedish corporation Pharmacia. He simply said that the decision to invest in Ireland was taken because the company wanted an environment similar to its own but with much less in the way of government regulation.[18] Attitudes like these help to create the climate in which accidental and even, in some cases, deliberate chemical releases can happen. In 1988 an Irish farmer finally won a case which had, on the way, bankrupted him, against Merck, Sharp and Dohme (a subsidiary of Merck US Inc.) whose factory on the banks of the River Suir had polluted farmland, killed cattle and produced devastating effects on local health for many years.[19]

Industrial chemists are perpetually creating new, unstable and poorly understood chemical substances which cannot be used, transported or disposed of with any confidence. Some of them, like PCBs, find their destructive way into our everyday lives with consequences which will affect not just us, but many generations to come. Other chemical combinations include 'me too' compounds designed to seize markets from their rivals. Some of these are relatively innocuous, as in the case of many non-prescription medicines, but others are not. Dangerous drugs are legion and their production and development must also be suspect, particularly in a world where the competition is so fierce. Roughly 11 million chemicals are now registered, of which about 100,000 are in commercial use throughout the world. About 1,500 of these make up some 95 per cent, in volume, of all chemicals in use and remarkably little is known about either their effect on the environment or their toxicity. In 1989 the OECD announced plans for launching a serious toxicological investigation into these chemicals. The first reports, due in 1993, are into a sub-group of commonly used substances and, at the time of the announcement, there were 147 of them, 70 of which were produced in annual quantities of 10,000 tonnes or more. By the time this part of the investigation is complete new, unpredictable and toxicologically unanalysed chemicals will be on the market.[20]

No 'treaty' on waste was constructed by the Global Forum during the Earth Summit, though there is a protest, in their document on nuclear power, at the unregulated and dangerous dumping of nuclear wastes on the South by Northern industries. Chapters 19 and 20 of *Agenda 21* both deal with the issues in relation to the developing world, but with that document's usual circumspection. The first of these is entitled 'Environmentally Sound Management of Toxic Chemicals Including Prevention of Illegal Traffic in Toxic and Dangerous Products' and it deals chiefly with the international trade in new chemicals. Obviously this includes the movement of dangerous chemical waste. Paragraph 55 describes the most serious of international worries:

> Many countries lack national systems to cope with chemical risks. Most countries lack scientific means of collecting evidence of misuse and of judging the impact of toxic chemicals on the environment, because of the difficulties involved in the detection of many problematic chemicals and systematically tracking their flow. Significant new uses are among the potential hazards to human health and the environment in developing countries. In several countries with systems in place there is an urgent need to make those systems more efficient.

The chapter points out that there is, at present, no international agreement governing traffic in dangerous products and that the only, somewhat flimsy, safeguards are embodied in various national laws which deal with illegal movements of hazardous chemicals. Both the United States and the Italian reactions, described above, when required to take back dangerous substances that they had originally exported demonstrate the inadequacy of national laws in the face of an international problem.

It is *Agenda 21's* Chapter 20, 'Environmentally Sound Management of Hazardous Wastes Including Prevention of Illegal International Traffic in Hazardous Wastes', which speaks most directly to the issue and it begins by recalling a resolution passed by the General Assembly on 22 December 1989, number 44/226:

> the General Assembly requested each regional commission ... to contribute to the prevention of the illegal traffic in toxic and dangerous products and wastes by monitoring and making regional assessments of that illegal traffic and its environmental and health implications. The Assembly also requested the regional commissions to interact among themselves and cooperate with the United Nations Environment Programme (UNEP), with a view to maintaining efficient and coordinated monitoring and assessment.

In paragraph 7 (b), (c) and (d) UNCED calls for the ratification of two important conventions: the Basel Convention on the Control of Transboundary Movement of Hazardous Wastes and their Disposal and the Bamoko Convention on the Ban on the Import into Africa and the Control of Transboundary Movement of Hazardous Wastes. It also demands of exporting countries that they at least stop exports to countries which have, either through the Bamoko or the fourth Lomé Conventions, prohibited the import of these wastes. Subsection A of this chapter suggests measures to promote the 'prevention and minimization of hazardous waste'. In essence this means campaigning for cleaner methods of production, setting goals for the stabilisation of the quantities of new hazardous wastes and 'promoting the use of regulatory and market mechanisms'. By the last of these they mean that governments should offer 'economic or regulatory incentives ... to stimulate ... cleaner production methods, to encourage industry to invest in preventive and/or recycling technologies' (paragraph 13 [b]).

Our summiteers did not go so far as to commit themselves to any action on these matters, but it is something that they agreed to their inclusion in *Agenda 21*. We shall have to see whether existing commercial practice will be modified enough to enable the developing countries efficiently to close their borders to waste-dumpers. Making that trade unprofitable could be the single most effective pressure towards less wasteful production.

9

Women and Other 'Groups'

Like all international treaties, statements of intent and similar instruments, *Agenda 21* is, in its structure, formulaic. Not only are principles, objectives, means and possible finances set out in a rigid order which makes it all very easy to follow, but virtually every chapter has a little mantra calling upon the gods or statespeople, whoever is the greater, to take into account 'groups that have, hitherto, often been excluded, such as women, youth, indigenous people and their communities and other local communities'. This example is taken from Chapter 10 on land resources. Even the smallest degree of textual criticism of the composition of the *Agenda* reveals, not surprisingly, a number of different authors. Nonetheless, few of them escape the charge of thinking of the young, of indigenous people, and, above all, of women as 'other groups'. We are entitled to ask 'other than what?'. All three, between them, make up most of humanity and the largest of these 'groups', women, means about half of the world's population.

This point may not be dismissed as an easy and fashionable political view. The continued habit of seeing women as 'other', as marginal, to be taken into account, even as a 'group', is to perpetuate the very problems to which the mantra was supposedly addressed. Women are not another 'group' whose special interests must be accommodated, they are half the world, and that means half of the young world and half of the world of indigenous people as well. It is depressing to find only a blurred and ineffectual recognition of this in the UNCED documents. What is at stake here is a whole way of looking at being, not a mere clumsiness of expression. So long as men continue to see themselves as the conscious centre and women as 'other', they will go on marginalising and commonly oppressing them. One further result is somehow to lump the multiplicity of women's interests together, yet they are as various as those of men and, not infrequently, the same. By making women a 'group', this variety is obscured.

No-one, apart from the lunatic far right, is interested in the perpetuation of the world's present inequalities and the consequent levels of deprivation. However, the world is not ruled according to some utopian rationality but by a vast number of sometimes interlocking and sometimes opposed special interest groups. These may be local, regional, national, international and transnational, private, governmental, or a mix of several of these. Each, with varying alliances, is determined to advance its interests and maintain its power and it is rare indeed for power to be surrendered voluntarily. It is from this kaleidoscopic arrangement of powers that women have largely been excluded and they cannot simply be grafted on because the ramifications of their exclusion are too extensive, too complex, in the end too cultural. Women, like the merchants of the Renaissance or the industrial workers of the early twentieth century, will have to seize power and in doing so will change the culture. Other writers and other books deal with most of the central issues of gender politics; our part is to see that these politics cannot be left out of any account of a sustainable and environmentally safe world.

Disparities between rich and poor countries bear most heavily on women and the inequalities from which women suffer in the developed world are exaggerated in the developing countries. Table 9.1 gives one crude measure, but it is important to remember that some inequalities simply affect women more than men. Poor land, inadequate and unsafe water, insufficient health-care, insecure food and energy supplies, poorer access even to simple technology and to markets, will all hit the health and the lives of women and the children that they must bear, first and hardest. Infant and maternal mortality, ill-health and the horrifying consequences of malnutrition are the results (see Table 9.2 for some indicators). These are the results, too, of a degraded environment and of unfettered free-market forces.

There is little in all this which is not widely acknowledged, but apart from a certain amount of supportive work from NGOs the customary Northern response is a gentle wringing of the hands in despair. Fortunately, Southern grassroots organisations are discovering new ways of attacking poverty and empowering the poor and so are becoming an increasing challenge to their own societies and to the world at large. Some concentrate on particular tasks, like the immensely successful Grameen Bank of Bangladesh which, on a unique principle of revolving debt among small cells of workers, makes small loans to the very poor who had always been spurned by conventional banks, development or otherwise. Others are more broadly directed, and among these few are more important than the growing number of women's groups. These often begin, particularly in rural areas, as informal groups of women organising the gathering of fuel or water, who then move on to other concerns. In Chapter 6 we offered two examples of women's groups which were growing from relatively informal structures into wider organisations, one in the Sahel and the other in West Bengal.

Table 9.1
Women's opportunities expressed as a percentage (men = 100%)

Area	*Developing countries*	*Industrial countries*
Unemployment	—	140
Primary education	91	—
Secondary education	70	—
Tertiary education	54	98
Third-level science education	—	62
Labour force participation	36	66
Representatives in parliament	15	22

Source: extracted from UNDP, *Human Development Report 1991*, Tables 10, pp. 138–9 and 31, p. 179, Oxford University Press, New York and Oxford 1991.

Another, major, example is the Self-Employed Women's Association (SEWA), which began in 1972 in Ahmedabad as a union of unlicensed women street-traders. In order to trade they had to go to money-lenders to borrow the cash with which to buy their produce, usually at extortionate rates of interest. They had also to bribe officials and police officers and to endure constant abuse and harassment. In combination and against vicious male and bureaucratic opposition, they fought the local state refusal to grant them trading licences, set up their own credit union and transformed their trading conditions. Since then they have moved into helping landless labourers to begin cattle-farming and have sponsored co-operative dairy businesses.[1] They continue, in these and in other activities, to keep the 'conscientisation', the defence and organisation of women, at the forefront of their activities and to this end have published a number of simple stories of their struggles which act as parables or exemplars for others.

Table 9.2

Mother and child survival and development

	All developing countries	*Least developed countries*	*Sub-Saharan countries*	*Industrial countries*
%age of births attended by health personnel	55	28	35	
%age low birth weight babies	18	22	15	
infant mortality rate per 1000 live births	74	115	106	13
%age under five underweight	35	45	33	
%age 12–23 months wasting	13	18	13	
%age 24–59 months stunting	40	52	46	
maternal mortality rate per 100,000 live births	420	740	690	26

Source: UNDP, *Human Development Report 1992*, Tables 11 and 12, pp. 148–51, Oxford University Press, New York and Oxford 1992.

Women's groups are as various as the needs they meet and they occur in urban areas as well as in those rural regions where women form the majority of subsistence farmers. In 1981 a group of women in Costa Rica formed CEFEMINA with the aim of improving the lives of urban women. One of its early campaigns was to compel a lethargic state to install the necessary infrastructure before building new housing projects. Subsequently it has campaigned, with some success, for undeveloped land to be handed over to community control so that self-help community development could be attempted and the natural environment preserved as far as possible and, most importantly, so that new settlements could be designed to improve the quality of life and to promote a sense of communal responsibility.[2]

Latin America provides many instances of grassroots women's groups, one of which is the Chilean urban women's group MOMUPO. Some sixty of its members combined with unions of domestic workers to create a literacy course which would attempt to 'stimulate creativity, critical understanding of reality and validate solidarity and social participation'. Basing themselves on the principles of Paulo Freire,[3] they recognised a 'new' literacy in television and in 1985 they created an immensely popular documentary video. Made by local women, it describes what happens to the lives of women who take part in organisations, it analyses the relations between women and men and it looks at the role of women in society.[4] They are discovering, as the Methodists of England and Wales discovered at the beginning of the twentieth century,[5] that literacy is an essential tool for emancipation.

Other kinds of Southern women's groups have become celebrated. One such was that great moment at the village of Saye in Burkina Faso when, after the men had talked for years of the need to build a dam to catch the run-off in the rainy season, the women, tired of waiting for action, built it. The Kenya Water for Health Organisation, made up of a group of NGOs, was a response to the self-help efforts of women to ensure safe water and sanitation. In Uttar Pradesh, in the village of Khirakot, women, on discovering that the state had given a contractor a mining licence and that his activities would prevent them from reaching their supplies of fuelwood and would destroy trees, banded together and compelled the courts to stop the mining.[6] There are many more such examples.

That women should be farmers, protect trees and reclaim their environments somehow fits with Northern liberal sentiment and the common, if sexist, view that it is all part of the image of women as mothers. That they should engage in finance, in urban planning, in fights for rights, in literacy programmes and for women's equality is altogether more threatening because these matters are the province of politics. It is particularly threatening to the surprisingly large number of Northern NGOs controlled by white middle-class men. Ritual nods are made in the direction of women's roles and even to the importance of their groups, yet nearly all practice continues not so much directly in defiance of women's demands but more as if they made not one jot of difference to some kind of objective reality.

All this is very odd, if only because throughout the Third World women have repeatedly demonstrated their ability to conduct affairs in ways which are sustainable exactly in the developmental forms described with such enthusiasm by these Northern men. That they can do so while also beginning to fight the war against a sexist world often means that they frequently develop social priorities which are different from those of most NGOs, as in the case of the campaigns for literacy in Central America. An excellent illustration of the point is to be seen in a grassroots NGO called the Centre for Self-Managed Development (CDA) set up by women in La Paz in 1983. It worked by bringing together a number of mothers' clubs which had sprung up among the Aymaran women in the *barrios* of El Alto and La Paz. The object was to reduce dependence on aid from external authority and replace it with self-management in order to address major problems neglected by the state. Infant mortality in El Alto, for example, was running at 208 in every 1,000 babies. Very quickly the group realised that in its fight against bureaucratic hostility and inertia it was engaged in a fight for the right of women to act in ways which might well challenge male priorities. Margarita Calisaya, one of the originators of the movement, said: 'We are not against men, but women are the axis ... the nucleus around which everything rotates. Women are the best means for disseminating knowledge and information to neighbours and to children.'[7]

It is because women's rights are so restricted – their rights to land, to what they grow, to organise, to make decisions – that CDA, SEWA and a host of other women's NGOs, in gradually becoming the means by which women are changing their situation, are also offering an edge of political hope. Their double battle against the restrictions of a male world as well as against poverty, while simultaneously often working as allies of men, makes them peculiarly inventive. They are the schools in which women the world over are learning how to seize the political initiative. However, they are not separate from other grassroots organisations but simply part of a continuum of Third World NGO activity which is emerging as a new and essentially political form of social organisation.

In very many instances the state is hostile to groups of this kind and it is no accident, for example, that many members of NGOs, particularly women, found themselves imprisoned in Pinochet's notorious stadium in Chile during the 1970s. 'Gabriela', a remarkable organisation of women in the Philippines which, with great courage, organised during the Marcos dictatorship, is at the time of writing still under constant threat from the military and the police. The women of SEWA perpetually face both domestic and state violence. This hostility is hardly to be wondered at, since these grassroots NGOs are in the business of wresting power over their own lives from ruling and frequently corrupt élites.

Despite the difficulty in their conceptual framework suffered by some of the draughtspeople of UNCED when trying to think about women, others were clearly pointing in more or less the right direction. Chapter 24 of *Agenda 21*, entitled 'Global Action for Women towards Sustainable and Equitable Development', in paragraph 3, urges governments to 'take active steps to implement', among other things, 'Measures ... to increase the proportion of women involved as decision makers, planners, managers, scientists and technical advisers in the design, development and implementation of policies and programmes for sustainable development'; and 'Measures to strengthen and empower women's bureaux, women's non-governmental organizations and women's groups in enhancing capacity-building for sustainable development'. Other measures called for in this paragraph include the promotion of better education for women, a reduction in their workloads, the promotion of female–male equality, the provision of access to good water and sanitation and to good fuel supplies, the creation of good health facilities and so on.

We have remarked before that there is a very strict limit to what we can expect from a United Nations conference and it is unquestionably a substantial advance for it to have achieved pious exhortation at least. Sadly, there is little evidence of these issues appearing as priorities in the national agendas of many, if any, of the participating governments. Yet the argument that runs through this book is that without them we are unlikely to find answers to the developmental and environmental ills of the world. Organisationally the keys lie in Southern and particularly in women's NGOs and in those Northern NGOs that have understood the matter. At present only they are able to make use of the slightly shaky endorsement national governments gave to these principles in Rio and only they are able to offer serious alternatives to the skewed and 'market' led programmes of so many of the developmental banks. We have, therefore, briefly to consider the phenomenon of NGOs and what we might expect from them.

Field-workers, researchers and activists of one kind or another within Northern NGOs have, in the last few years, increasingly recognised that their role is no longer so straightforward as it once may have seemed. The variety of grassroots organisations which have emerged in the South, particularly in the last few years, raises acute questions of identity and purpose for the organisations of the North. As people in developing countries are producing their own organisational responses to poverty and finding their own ways of analysing and meeting needs, they call into question the place of the Northern groups. It is to the credit of many that individuals within Northern NGOs have responded well to the challenge, and that numbers of NGOs are formally examining the issue, but they are hampered by a complex of structural problems.

These arise from political and economic differences which we may categorise fairly schematically. The developed world in general has strong civil societies with multi-party governments; it has strong traditions of organised labour and, as a rule, advanced welfare systems with a wide level of entitlement. Northerners treat the environment largely as something to be consumed in leisure and to be 'fixed' when nature rebels. In one form or another Northern citizens invest in insurance against natural phenomena and are prepared to make charitable donations to, or engage in voluntary work for, those less fortunate than themselves. Their economies and industries generate capital for investment and reserves and they do so from internal resources. A strong management culture exists together with institutionally determined sets of priorities administered by line management. The creation and use of technology is a normal part of the production process. Northern states largely define their own national agendas and their citizens have at least the illusion, if not always the reality, of participating in those definitions.

In the Third World all this is entirely reversed. Civil society is weak and government is commonly single-party; there is less in the way of traditional organised labour and very weak welfare provision. The environment and its resources are a basic need and nature is often a threat whose control must be limited. Large families are the only insurance against risk and Northern 'voluntarism' becomes patronage in the South. Capital generation is replaced by revenue expenditure and there is little, if any, investment; such funds as there are come from negotiation with external bodies. In place of management there is a culture of command, priorities tend to be set by the commanders and group negotiation takes the place of line management. Any technology is normally imported. Third World states must respond to internationally determined agendas rather than help to create them. We may express this in tabular form (see Table 9.3).

Table 9.3

The socio-economic context for NGO development

First World	*Third World*
Strong civil society	Weak civil society
Multi-party	Single party
Strong tradition of organised labour	Weak tradition of organised labour
Strong entitlement to welfare	Weak entitlement to welfare
Environment as a leisure Consumption activity	Environment as a basic need
Controlling the limits of nature	Limiting the control of nature
Capital insurance against risk	Extended families to minimize natural phenomena
Voluntarism	Patronage
Increasing capital saving	Increasing revenue expenditure
Funds internally generated	Funds externally negotiated
Management culture	Caudillo culture
Institutionally set priorities	Personally set priorities
Line management	Group negotiation
Technology part of production process	Technology an imported product
Involved in defining national agendas	Responds to internationally set agendas

Source: Ian Cherrett, ETC UK.

These differences underlie NGO structures in both the developed and the developing worlds, though in importantly differing ways. Northern organisations reflect the confidence of their industrial worlds; they tend to be professional, to have well-organised methods of fundraising and a management approach to problem-solving. No matter how far along the road they may go towards participatory action in the field, they cannot help but be interventionist and Southern perceptions of this are heightened by the frequent NGO involvement with governments.

In the South the picture is more complicated and there are two categories of NGO, which may broadly be divided into grassroots and professional. Grassroots organisations are usually community-based, although they may be federated, and they concentrate on the process by which their aims might be fulfilled; they also tend themselves to be the centres around which local civil society coalesces. Membership, usually untrained, is drawn from the marginalised poor, shares the local culture and, because it lives there, is based in the field. Their decision-making is normally communal, their key people are activists and ideology is forged through practice. They are legitimised through successful battles and are crucial agents in the formation of democratic and developed societies.

Professional Southern development NGOs tend to be a response either to government or to donor agencies. They are usually constructed to do a particular job and they reflect the working structures of the civil society from which they spring. Their culture tends to be international, they are highly trained, not in any sense representative of the people among whom they work, and generally middle-class. Projects are normally individually and externally funded and their policies are made by boards of directors who are, as a rule, answerable to, or at least heavily swayed by the ideas of, the funding source. Decision-making is centralised and ideology is received. Their centre is their headquarters and so they have no local base. Their legitimacy lies entirely in their success in raising the funds and expertise to fulfil particular projects.

NGOs working in the fields of development and environment are usually charities and they share an ethos with all other charitable bodies which, to a considerable degree, conditions the ways in which they fulfil their purposes. Philanthropy, whether or not embodied in institutions, has a dubious history if only because it is so readily transmuted into an expression of power. To give in order to relieve want in whatever form may be generous, but the act rarely, if ever, addresses causes and in failing to do so often runs the danger of reinforcing them. We have already remarked that early capitalist states, to go no further back in history, by striving to limit questions to do with their citizens' welfare to issues of control, set precedents which still condition much social thinking. Charities, largely religious and frequently based on the corrupted Calvinist or post-Tridentine theologies, provided for the worthy needy. 'Worthy' meant consenting to the mores of church and state and accepting at last that:

The rich man in his castle,
 The poor man at his gate,
God made them, high or lowly,
 And order'd their estate.[8]

Charity began not at home but in the corridors of power, and despite the generosity of many individuals it was directed towards keeping the lower classes in order. In Britain, as in much of Europe (consider Zola's *Germinal*), it took the rise of trade unionism, the emergence of working-class literacy and the building of the co-operative movement to wrest some of those mechanisms of control from the hands of the ruling élites.[9]

The connection between philanthropy and the worthiness of its objects has been hard to shake off. It crops up, for example, in the iconography of many of the appeals made by NGOs; the granny to be adopted dresses demurely, smiles engagingly at us and generally looks like someone we would not mind having around the house; the diseases of want strike the pretty child; a modest donation provides a hard-working peasant-farmer with a pump for irrigation. Charities are in a cleft stick: if they portray reality they frighten the donors and get accused of exploiting misery or purveying despair, but if they use bright hopeful images they reduce the power of the people empicted to challenge Northern liberal assumptions. The appearance, at least, of the poor must conform to middle-class standards of decency.

For Northern NGOs involved in issues of aid and development in the Third World, the object is to make possible standards of living which will bear comparison with those of the North. In many cases this involves working on quantities of individual projects of varying size which will enable the poor in each of the places where the projects are undertaken to improve their immediate economic circumstances in sustainable ways. Deep in the collective unconscious of NGOs is the hope that success in a project will set a good example; if neighbouring villages, regions, even countries see that something has worked well, they may be persuaded to follow suit. By this means will sustainable development spread and, by a process of 'trickle-up', economies may be transformed. It is a generous illusion and, although most individual members of these organisations are not so ill informed as to let it take over their conscious lives, it comes from the liberal charitable origins of their movements and is one in which the seeds of failure lie. Northern charity has always been ameliorative, a bandage for the wounded while change is being fought for elsewhere. This background makes it very difficult for Northern NGOs to offer challenges to the fundamental political and structural problems that are the first causes of the poverty that they are addressing.

If we are right about this then it is not surprising that the culture from which Northern field-workers come inclines them to developmental and economic models which are mimetic of the richer world, even though in some cases they may be engaged in trying to iron out that world's 'errors'. There is a sense in which even the most 'scientific' of economic, industrial and agricultural processes is ideologically conditioned. At basic levels where the object is, say, sustainable subsistence farming and land reclamation it is not uncommon, for instance, to find that charitable aid groups are happier working with men than women, or in societies where there are settled people and recognisable patterns of land tenure. Many religious NGOs like the LWF or the Catholic Institute of International Relations (CIIR) sometimes work in surprisingly radical ways and do so as an expression of their Christian commitment to their fellow beings rather than from any directly missionary purpose, but the missionary hope is never entirely absent; it is no accident that the CIIR, for example, was once called 'the Sword of the Spirit' and had the extension of the gospel as part of its aims.

Some of the most influential NGOs, like the British Oxfam and the United States' CARE (Co-operative Agency for Relief Everywhere), began their existence as relief organisations either during or following the two great Northern wars and gradually turned their attention to activities in the developing world as Northern need declined. Their initial policies were devoted to trying to get the poor of the South sufficiently economically active to be able to reproduce the social and economic patterns of the North. During this period, which came to an end in the 1960s, Oxfam was not alone in channelling its funds through Christian missionary organisations; European states, particularly Britain, were compromised by their colonising past and missionaries were thought to be the nearest thing to independent agents.

Since the 1960s much has changed. Most Northern NGOs have learnt the value of participatory enterprise, have learnt to listen to the people they are aiming to assist and have realised that the South has its own valid views about the paths to development. Perhaps the greatest change of all has been the growing acknowledgement, particularly among NGO field-workers, of the political, financial and industrial bases of unequal development. This has brought many organisations to a difficult place. Some states, Britain in particular through its policing agency, the Charity Commissioners, have draconian rules governing what charitable bodies may or may not do. Chief among these is a fairly stringent prohibition on anything that smacks of political activity and this is invariably interpreted (certainly by the Charity Commissioners) as anything which, in some way, might upset the views of Tory back-benchers or of M. le Pen. In the forlorn hope of countering state censorship, some NGOs have begun to call part of their activity 'advocacy' and there are even groups, like Health Action International who campaign for the international control of pharmaceuticals, which engage in nothing else. Another stratagem has been to eschew party politics and claim that NGOs are neither of the right or the left, but are simply radical. This particular liberal chimera may serve to keep the state's thought police at bay, but it is too weak a position to merit attention.

Advocacy from those organisations working among Third World communities must consist of providing the poor with a voice in the North and, as that voice cannot simply be of the begging bowl, it must be political – once again the instance of Oxfam's attack on apartheid springs to mind. So among the major challenges facing Northern NGOs is their growing need to fight on the political front. This is simply because the relationships between North and South are relationships of power and control. NGOs are not yet able completely to escape the accusation of being instruments of Northern domination, partly because they are often conduits for Northern aid money (see Table 9.4) and partly because they do not, as a rule, directly address the political problem of Northern hegemony. Perhaps the most notorious example of this is to be seen in the way that the United States government aid ministry, USAID, has influenced the shape and progress of United States NGOs, frequently turning them into instruments of its foreign policy. Canada, too, through its state agency CIDA (Canadian International Development Agency), has a dispiriting record of only recognising those grass-roots organisations which will carry out its policies. One of CIDA's more dismal moments, for instance, was the over-riding of Barabaig land rights in its pursuit of sustainable practices for a neighbouring group.[10]

Table 9.4
Comparative figures for official and for NGO aid (US$ billions)

Years	*70*	*75*	*80*	*82*	*83*	*84*	*85*	*86*	*87*	*88*
OECD aid	7.0	14.3	27.7	28.2	28.1	28.7	29.4	36.5	41.6	48.1
at '86 prices	22.2	26.0	31.7	34.2	34.2	35.6	36.1	36.5	36.0	38.7
% increase		17.0	22.0	7.9	0.03	4.1	1.5	1.1	-1.4	7.5
NGO aid*	0.86	1.35	2.39	2.32	2.32	2.6	2.88	3.34	4.0	4.2
as % of GNP	.041	.034	.032	.029	.029	.038	.041	.035	0.3	0.3
at '86 prices	2.73	2.44	2.73	2.81	2.82	3.22	3.54	3.34	3.1	3.4
% increase		-10.6	12.0	2.7	0.4	14.5	9.8	-5.7	-7.2	9.7
NGO total †	2.77	2.64	3.93	4.21	4.25	4.86	5.35	5.29	4.7	5.2
% increase	-4.7	-4.7	48.7	7.3	1.0	14.4	9.9	-1.1	-11.0	10.6

Notes: * Private donations. † 1986 prices includes official aid through NGOs.

Source: John Clarke, *Democratizing Development*, Earthscan Publications Ltd, London 1991.

It is not only as surrogates for state funding that Northern NGOs face some incredulity from the South. Many of them wield huge budgets: the United States agency, Catholic Relief Services, has a budget which exceeds that of Belgium and Oxfam's is greater than that of New Zealand. They are also large-scale employers and must have the care of their Northern employees in mind; this, too, makes them less likely to take ideological risks. Although the Club of Rome estimates that Northern NGOs work with between 10,000 and 20,000 Southern counterparts,[11] there are probably equally substantial numbers remaining outside the charmed circle. Clues to the instincts which dictate the policies of partnership are to be found in the cultural and structural differences in the development of Northern and Southern NGOs and also in the differences in Southern responses to Northern intervention.

These differences are not capable of simple resolution. Northern NGOs work with large numbers of Southern groups and they are able to help with the necessary financing of them. This, in turn, produces the danger of creating a system of three classes, made up of Northern 'experts' and relief workers, Southern NGOs partly supported by the North, and grass-roots organisations which battle on alone; but where Northern NGOs are aware of this they may sometimes circumvent it. The dilemma arises because Northern NGOs cannot solve the problems, and at one level they are part of them, both as agents of the North and even, as in the case of their frequently internally sexist organisations, as the embodiment of Northern values.

While this issue cannot be evaded, it does not amount to a condemnation. For as long as it is politically possible, NGOs must continue to occupy that uneasy position of campaigning in the North for money, agreements and sensible policies and supporting, so far as they can, groups which are subverting the Northern order of things. This has become particularly true since June 1992, when governments at Rio approved the increase in funding for the Global Environment Facility from US$1.3 billion to around $3 billion. That money is likely to be administered by the World Bank, that most suspect of all development organisations.[12] Many Northern NGOs campaign hard for the principle that developmental and environmental problems will be tackled successfully if only the poor are empowered; it must be their final role to apply that principle to themselves and surrender their position to Southern groups. A significant step in that process would be to urge the shift of such enormous financial power from the grasp of an agency so firmly wedded to the developmentally and environmentally destructive ways of Northern industry as the World Bank and into some less compromised agency.

Despite the rapid increase in the use of NGOs as the means by which governments deal with the South, they were not asked to participate directly in the Earth Summit, but only to appear as advisers. The message was, however, a little confused simply because so many industrial groups were recognised as NGOs. Probably the most important point to be gathered from this fiasco is that UNCED was simply a reversion to the 1950s' and 1960s' model of development which saw the incorporation of the South into the world market as the answer. The collapse of the market into world-wide recession accompanying the collapse of the centralised command economies seems to have passed the Summit by.

NGOs had to fight quite hard even to be allowed to present their views to the preparatory committees of the Summit, largely because the delegates of some governments, including a few from the South, insisted on a legalistic interpretation of the UN resolution (General Assembly 44/228) which restricted NGOs to 'consultative status with the Economic and Social Council'. NGOs were also excluded from any of the preparatory negotiations about the nature of the agenda. Only on the occasions when the committees turned to more general discussions could the NGOs deliver positional papers. All the substantive negotiations at the Earth Summit, when it finally took place, were conducted in private and outside the plenary sessions. NGOs were not party to them. UNCED was charged, in theory, with examining progress since the presentation of the Brundtland Report – the irony of all this will not be lost on those who have read the report and come across the words:

> NGOs and private community groups can often provide an efficient and effective alternative to public agencies in the delivery of programmes and projects. Moreover, they can sometimes reach target groups that public agencies cannot ... governments, foundations and industry should also greatly extend their co-operation with NGOs ... To this end, governments should establish or strengthen procedures for official consultation and more meaningful participation by NGOs in all relevant intergovernmental organisations.[13]

Crumbs of comfort may be found in Chapter 27 of *Agenda 21*, 'Strengthening the Role of Non-governmental Organizations: Partners for Sustainable Development'. It is among the shorter chapters in the Agenda and its first paragraph begins unexceptionably:

> Non-governmental organizations play a vital role in the shaping and implementation of participatory democracy. Their credibility lies in the responsible and constructive role they play in society. Formal and informal organizations, as well as grass-roots movements, should be recognized as partners in the implementation of Agenda 21.

What the chapter goes on to say is that NGOs 'will be of particular importance to the implementation and review of environmentally sound and socially responsible sustainable development'. It is careful not to particularise and contents itself with telling governments that they ought to make use of all that NGO expertise, that they should establish mechanisms of one sort or another and review their levels of support. Good solid stuff and exactly what the developing world needs.

We remarked in Chapter 1 that many Northern governments might well see donations to NGOs as a surrogate for taxation. Aid budgets are an easy target during a major slump and as they are slashed that voluntary 'tax', together with NGO expertise and field capacity, may well become the widely accepted and cheaper alternative. This may not altogether be a bad thing. NGOs do not, as a rule, engage in investment designed to achieve a handsome and repatriable profit. But the best of all things that we may hope to see arise from a substantial reduction in Northern state activity on the aid front is a far larger role for the NGOs as the administrators, the instruments, of Northern aid policies. Given the unquestionable goodwill of the Northern NGO movement, this change could lead to a healthy polarisation between Northern and Southern NGOs and the consequent increase of Southern grass-roots movements designed to replace the Northern charities.

10

Debt is Bad for Your Health

As we pointed out in Chapter 1, sustainable development was defined, in some senses quite usefully, in *Our Common Future* as 'development that meets the needs of the present without compromising the ability of future generations to meet their own needs'. Since its publication the phrase has become something of a cliché, often without its devotees quite recognising that, on its own, it is largely devoid of meaning. It is not too large a generalisation to say that most of us fill out the phrase with suitable agricultural or village images, and with some reason. As we have said before, although the picture is changing very fast, most people still do live in rural areas. There is also a sense in which, gigantic as the task may be, rural development, to a level where far fewer people are malnourished, seems feasible. Even allowing for the unparalleled rural catastrophes unravelling in Africa at the moment, working for food security based on sustainable farming methods, both subsistence and cash-crop, seems to demand fewer resources for a larger return than any other course of action. Because the techniques are relatively so simple and the results usually so impressive, the goals of sustainable agriculture and arboriculture permit all those involved in aid and development to set themselves what seem to be practical and practicable tasks. This allows many people to put the structural causes of unsustainability to one side. Grumbles may be heard, but the big questions are too big and are for others, or they are 'cross-sectoral' and outside the immediate remit of the Northern agencies involved, or they are 'political' and so depend on larger world processes for their solution.

So far in virtually every chapter of this book we have made the point that UNCED also failed to come to terms with these structural causes. This is partly because faced with apparently intractable problems the bureaucrats opted for what they considered to be the least contentious approach, and hence the one most likely to succeed. Development had to be seen through the lens of environmental distress, and only the flat-earthers among the delegates would doubt the evidence of widespread destruction. We have already remarked that the agenda for the Summit was almost exclusively environmental. If, however, we were to be generous, then we must accept that the organisers really did believe that in the process of righting environmental damage, a process plainly in everyone's interest, the developmental difficulties which compound the problems must be put right.

We are in the middle of *fin de siècle* economic catastrophe in which capitalism, or, as it prefers to be known, the 'free-market economy', is in a bad way and an apparently unstoppable recession is running wild throughout much of the Northern world. Stalinism, in finally leading to the collapse of Third International communism, had succeeded where the Wars of Intervention and Operation Barbarossa had failed. Deplorable as the Soviet empire may have been, its disappearance has somehow added to political and economic insecurity. The great Northern slump of the 1930s disappeared into the Second Northern War, and it seems quite possible that massive investment in conducting the Cold War staved off the economic mess in which the North now finds itself.

We have no reason to be generous. UNCED was little more than the response of a frightened North which is in the process of turning its back on the poor of the world. Instead of recognising the importance of at least trying to find solutions to misery, the North is busy securing its immigration laws against the South. In future refugees may be admitted if they can conform to some arcane definition of 'political', but will be refused if they are merely 'environmental' or 'economic'. In this the North has bowed to its new neo-fascist paranoiacs. Development funding will increasingly be directed not at the developing world but at the incorporation into a loose Northern alliance of what can be rescued from Eastern Europe and from the fragmented Soviet Union.

According to the Brundtland Report, in 1980 there were 340 million people, spread through 87 countries, who were not being fed well enough to avoid stunted growth and serious risks to their health from the diseases of acute poverty and malnutrition (see Table 9.2). It was the World Bank's view that this figure would continue to grow.[1] Accurate information is hard to come by, but a survey conducted by the United Nation's Children Fund (UNICEF) in 1990 pointed out that, excluding China, one-third of the world's children, that is about 150 million, below five years of age are malnourished. Sixty per cent of these under-fed children are in Asia, and in sub-Saharan Africa the numbers also seem to be growing. If the figures for China are included, then 80 per cent of the malnourished children are in Asia.[2] Malnutrition in children is commonly an early warning of incipient famine and is certainly an indicator of widespread malnutrition among people of all ages within the communities in which it occurs. The number of those living barely above the extreme danger level is legion and it is clear from the UNICEF figures, which are for children alone, that the situation has deteriorated dramatically since the 1980 figure given in the Brundtland Report.

Another measure of the gravity of the situation is to be seen in the increasing income disparities between the richest 20 per cent of the world's nations, who receive 82.7 per cent of global income, and the poorest 20 per cent, who get a miserable 1.4 per cent (see Table 10.1). There can, of course, be enormous differences within poor states (indeed, within rich states too) and in this, as in so many sadnesses, Brazil leads the way. In that country the richest 20 per cent of the population has an average income some twenty-six times that of the poorest 20 per cent. If world-wide disparities of this kind could be more thoroughly analysed, then it has been estimated that the 59–1

Table 10.1
Global income disparity, 1960–1989
(percentage of global income)

Year	*Poorest 20% (%age)*	*Richest 20% (%age)*	*Richest to poorest*	*Gini* coefficient*
1960	2.3	70.2	30–1	0.69
1970	2.3	73.9	32–1	0.71
1980	1.7	76.3	45–1	0.79
1989	1.4	82.7	59–1	0.87

*A statistical measure of inequality, in this case on a scale where 0 = equality and 1.0 = total inequality.

Source: UNDP, *Human Development Report 1992*, p. 36, Oxford University Press, New York and Oxford 1992.

ratio, given in Table 9.1, could easily become an almost unimaginable ratio of 150–1. To get to that figure the comparison would have to be not between countries, but between the richest and the poorest 20 per cent of the world's people. Unfortunately analyses of this kind can only ever be approximations because so many countries do not publish their figures. The gaps are getting larger and although it may not be possible to produce the overall statistical ratios, estimates do exist for the difference in absolute terms between the incomes of the richest and the poorest 20 per cent of people: the gap has grown from US$1,864 in 1960 to $15,149 in 1989.[3] As the World Bank figures show (see Table 5.1), the situation will get worse because among the poorer states GNP is growing far too slowly, and in some instances it is actually declining.

As the poor get poorer, at least in part because they are paying for the unregulated growth of the North, not only is the environment degraded in ways on which we have touched, but the 'human capital' of the South is also being squandered. Change in the South will not only depend on less destructive patterns of Northern behaviour and on the transfer of assets and technology, but also in building the infrastructures of health and education, which are at present not merely deficient but actually weakening. If people are not only hungry but diseased and inadequately cared for and if there is no widespread culture of education to tertiary levels, then many of the paths to development are closed. Indicators of the sizes of the differences in these two fields between the North and the South are to be seen in Tables 10.2 and 10.3. In the case of education the smallness of the intake into secondary and tertiary education becomes self-reinforcing simply because the skills and expertise which are the essential basis for any extension of education are just not there. Neither, of course, can there be enough of the qualified people needed for modern industrial and scientific research and development. As a means of coping with some of these differences UNICEF has proposed a 'debt-for-child-development swap' in which debt could be redeemed in exchange for a commitment from the debtor to devote the funds to child development, including education. The cause may be good, but the arguments against these debt swaps remain. In any case that such an idea should be widely adopted seems to us improbable because, unlike the debt-for-nature swaps, there is too litle to be gained from it by the North.

Table 10.2
Access to health services/expenditure

	% of population with access to health services 1987–89	*Population per doctor 1984*	*Population per nurse 1984*	*Public expenditure on health (as %age of GNP)*	
				1960	*1987*
All developing countries	64	4,590	1,910	1.0	1.4
Least developed countries	46	21,410	4,910	0.7	1.0
Sub-Saharan Africa	48	22,930	2,670	0.7	1.0
Industrial countries	—	460	150	—	8.3

Source: extracted from UNDP, *Human Development Report 1992*, Table 12, pp. 150–1, Oxford University Press, New York and Oxford 1992.

We have touched briefly on a number of the consequences of these disparities and, at greater length in the case of trade, on the causes. But behind all the ills of inequity lies the matter of debt. If World Bank estimates are to be trusted then the total external debt of the developing countries now stands at the ludicrous figure of around US$1,306 billion. In absolute terms Brazil is the country which owes the most, though obviously the seriousness of debt depends on the value of production within the debtor economy. Table 10.4 charts the progress of real GDP rates per capita and those with the most dismal record are, as usual, the most deprived. To give a further indication of what is happening, we give a league table of those developing countries which owe $10 billion or more and we express these sums as a percentage of their GNPs (Table 10.5). The crude result of debt-servicing is that while in 1980–2 cash flows from the rich to the poor countries were in the region of $49 billion, in 1983–9 the situation was reversed and the poor paid over to the rich a net sum of $242 billion. As we pointed out in Chapter 2 , as much as half of the export earnings of many of the poorest countries go to debt-servicing. It is also notorious that many of these debts were incurred by corrupt leaderships suborned by banks anxious to find ways of making returns on, among other funds, their excess petro-dollars. Susan George is, perhaps, the best known of those who have written eloquently of the injustice of the crippling debts under which the developing world is labouring. She has also made the point that interest payments exported to the lenders have, over the years, wildly exceeded the totals of the original loans.[4] This sort of lending is known, in the poorer quarters of Britain's cities, as 'loan-sharking'.

Table 10.3

Education enrolment ratios, 1988–89

	Primary enrolment ratio		*Secondary enrolment ratio*		*Tertiary enrolment ratio*	
	total	*female*	*total*	*female*	*total*	*female*
All developing countries	94	80	41	35	8	5
LDCs	65	54	45	16	2	1
Sub-Saharan Africa	67	60	41	17	2	1
Industrial countries	—	—	85	86	37	35

Source: extracted from UNDP, *Human Development Report 1992*, Tables 14 and 34, pp. 154–5 and 193, Oxford University Press, New York and Oxford 1992.

That these debts make it virtually impossible for poor nations to invest properly in energy, industry, housing, health and education is one massive penalty. Another is the degree to which they skew under-developed economies. Production of any kind to satisfy an export market has become predominant, with, as we have pointed out before, the consequence that in the agricultural sector farmland has increasingly been turned over to export crops which depend on high chemical inputs on over-used land from which traditional subsistence farmers have been driven. The plight of poor farmers (half of whom, as we have frequently said, are women) who have thus been marginalised and whose land is often consequently degraded, becomes yet more desperate. In Southern industrial sectors production is also largely directed to the export of raw materials while the manufacture of processed goods in many parts of the developing world, as we observed in Chapter 5 , is determined by the restrictions of Northern trade barriers. Debt-induced poverty leads directly to conditions and pay in the mines and factories of the developing countries which are frequently inhuman; bonded labour, particularly of children and women, is common, as it is in many rural areas, and it is often difficult to distinguish it from slavery. Several transnational corporations are among the contemporary slave-owners.

Table 10.4
Developing countries – real GDP per capita by region*
(1970 = 100)

Year	*70*	*72*	*74*	*76*	*78*	*80*	*82*	*84*	*86*	*88*	*90*	*92*
Africa	100	105	110	115	110	110	108	100	100	100	100	100
Asia**	100	110	115	120	135	140	140	150	180	190	200	250
Europe	100	105	120	140	145	143	141	150	160	161	155	160
Mid. East	100	110	130	135	140	140	120	110	100	105	95	95
W. hemisphere	100	105	120	125	130	138	122	125	127	125	120	122
Asia												
Hong Kong, Indonesia, Korea, Malaysia, Singapore, Taiwan, Thailand	100	120	130	140	160	170	190	210	220	260	310	320
China	100	110	120	120	140	155	170	200	240	280	310	330
Rest of Asia	100	105	105	110	120	120	125	130	135	145	155	160

* Composites are average figures for individual countries weighted by the average US dollar value of their respective GDPs over the preceding three years (1992 figures are staff projections).
** This line is analysed in the lower Asian section.
Source: IMF, *World Economic Outlook 1991*, p. 13, Washington DC 1991.

In the run-up to UNCED even the redoubtable Mostafa Tolba, the head of UNEP, is reported to have said: 'We are convinced that the best way to achieve environmentally sound and sustainable development is by relying on proven tools – the free market, a climate friendly to investment and our God-given ingenuity.'[5] Ingenuity may be a gift from God, but 'free' markets and climates of investment come from the more dubious hand of Mammon. We are entitled to ask if Tolba is quite happy with debt as well. In a depressingly chirpy report issued by UNEP UK and the International Institute for Environment and Development (IIED), some months before Rio and shortly after the meetings of UNCED's third preparatory committee, it was remarked that the preparatory committee saw debt as a cross-sectoral issue and hence 'more to do with causes than sector by sector problems'. So, the report goes on, 'it is not surprising that the Prepcom [*sic*] failed to grapple' with it. In the event the Earth Summit neatly sidestepped the question; the conference was environmental and aimed at getting international agreements on environmental and ecological problems, so it could safely ignore debt. But because it intended that the issues should be pervaded by a developmental dimension, its documents do make occasional genuflections in the direction of debt elimination, and we must turn again to *Agenda 21* for an account of the Summit thinking on the matter.

We suppose that the absence of a chapter specifically devoted to debt in the *Agenda* is either for reasons that we have just set out or because the potato was simply too hot to handle. Nonetheless debt is actually mentioned in four of the forty chapters and in the 'Non-legally Binding Authoritative Statement of Principles' on forests. There is a sentence in the last of these, so opaque as almost to defy analysis, that brings together three of the most durable elements in developmental problems. It speaks of the need those countries have for support in trying to improve their forest management. Any party offering that support, the sentence continues, should take account of the 'importance of redressing external indebtedness, particularly where aggravated by the net transfer of resources to developed countries'. Through a tangle of sub-clauses, the sentence lurches to an end saying that Southern states need to achieve, at the very least, the replacement value of their forests and that they really should be allowed to market their forests products, particularly processed forest products, freely. Just so, if not exactly in a nut-shell, there it all is: debt and consequent poverty, environmental degradation (because that follows any failure to achieve 'replacement value') and the restrictiveness of Northern trade agreements.

Table 10.5
Developing world's most deeply indebted countries/debt as a percentage of GNP

Country	*Total debt (US$ millions)*	*as %age of GNP*
*Upper-middle-income economies**		
Brazil	116,173	24
Mexico	96,810	51
Korea	38,014	16
Venezuela	33,305	80
Middle- and lower-middle-income economies		
Argentina	61,144	120
Turkey	49,149	54
Philippines	30,456	65
Algeria	26,806	57
Thailand	25,868	34
Morocco	23,524	98
Peru	21,105	74
Malaysia	19,502	52
Chile	19,114	78
Côte d'Ivoire	17,956	182
Colombia	17,241	46
Syrian Arab Republic	16,446	47
Ecuador	12,105	117
Nicaragua	10,497	370†
Low-income economies		
India	70,115	24
Indonesia	67,908	59
China	52,555	11
Egypt	39,885	159
Nigeria	36,068	119
Pakistan	20,683	47
Sudan	15,383	71
Bangladesh	12,245	53
Zaïre	10,115	97

*These divisions are World Bank categories.

†Nicaragua's GNP is uncertain, this figure is based on a World Bank estimate.

Source: compiled from World Bank, *World Development Report 1992*, Table 21, pp. 258–9 and UNDP, *Human Development Report 1992*, Table 19, pp. 164–5, both Oxford University Press, New York and Oxford 1992.

Textual criticism is really a futile way of gathering clues about the intentions of politicians; what they do is usually a more reliable guide to their intent than what they say. It would be rash, therefore, to put too much hope in the title and approach of Chapter 2, the first substantive chapter in *Agenda 21.* It bears the title 'International Cooperation to Accelerate Sustainable Development in Developing Countries and Related Domestic Policies' and is quite robust:

> The development process will not gather momentum if the global economy lacks dynamism and stability and is beset with uncertainties. Neither will it gather momentum if the developing countries are weighted down by external indebtedness, if development finance is inadequate, if barriers restrict access to markets and if commodity prices and the terms of trade of developing countries remain depressed. The record of the 1980s was essentially negative on each of these counts and needs to be reversed. (paragraph two)

The demand is repeated in paragraph 3 and returned to yet again in paragraphs 23 and 24. In the second of these two we find what is possibly the strongest statement in the entire *Agenda:*

> the reactivation of development will not take place without an early and durable solution to the problems of external indebtedness ... The burden of debt-service payments on those countries has imposed severe constraints on their ability to accelerate growth and eradicate poverty and has led to a contraction in imports, investment and consumption. External indebtedness has emerged as a main factor in the economic stalemate in the developing countries. Continued vigorous implementation of the evolving international debt strategy is aimed at restoring debtor countries' external financial viability, and the resumption of their growth and development would assist in achieving sustainable growth and development.

We should cling to any hope that we may derive from this sign that the message, so long trumpeted, is getting through, even when the response is confined to that flatulent recommendation that 'governments' should 'take account' of these things.

What seems to lie behind paragraph 24 is that good old Keynesian principle that the way out of a recessive economy is to spend. If, the *Agenda* seems to be saying, the North promotes the 'growth and development' of the South, presumably financially as well as in other ways, then that would 'assist in achieving sustainable growth and development'. Southern growth might well assist the ailing economies of the North, simply by broadening the efficiently trading community. Much of the North has offered its own, inimitable, response to this apparently sensible suggestion: it is cutting aid budgets and (in the GATT and in the NAFTA) erecting further hurdles for Southern traders. We may look for an explanation of this plainly aberrant behaviour in that curious state sectoralism:

> Officials are highly educated, but one-sided; in his own department an official can grasp whole trains of thought from a single word, but let him have something from another department explained to him by the hour, he may nod politely, but he won't understand a word of it.[6]

A generalised failure among Northern 'officials' to associate effects with causes, or to understand the other department, is best illustrated by that curiously named monetarist stratagem, 'structural adjustment', so adored by those terrible twins, the World Bank and the IMF. Forced upon governments fatally weakened by debt, structural adjustment means the drastic reduction of welfare among their already dreadfully deprived people, the privatisation of most of their few remaining state assets, the removal of normal financial controls and the opening of their markets to foreign corporations. Neither of the two agencies are in the least put out by the complete absence of any evidence that such measures have ever worked; after all what is at issue here is not reason, but faith, some esoteric and brutal rite of initiation into the sacred mysteries of the 'free' market. To pay for this even greater onslaught on their economies, many countries, particularly in Latin America during the 1980s, were compelled to yet further and even more unsustainable and ecologically devastating depredations on their dwindling natural resources. Not only in Brazil but throughout Amazonia, for example, drilling for oil and open-cast mining destroyed the rainforests every bit as fast as the hamburger connection. Faith can not only move mountains, it can hack down forests, pollute rivers and impoverish people.

Poor countries frequently have weak democracies, sometimes as a result of inherent instabilities which are at least enhanced, if not caused, by poverty. But even without that difficulty, the developing world was caught up in the lunacies of the Cold War as individual countries, or even blocs of countries, were seen by the contestants in that war as strategically useful. Instability was increased by the number of Southern states ruled by puppet dictators installed by one or other of the 'great powers'; Guatemala, North Korea, Uganda, the Philippines and Laos are all well-known examples. Wars of intervention, like those in Mozambique, Angola, Viet Nam and Afghanistan, were also common. Occasionally small adventurist excursions by Northern states occur: the British expedition to the Malvinas Islands, sometimes quaintly referred to as 'the Falklands War', was one such; the United States' occupation of Grenada another, and the Soviet invasion of Afghanistan, which so faithfully reproduced the British attempt of 1878–80 in the same country, was a third.

A cynic from some other planet might be excused for observing that all this leads to substantial arms expenditure which, in turn, both profits Northern industry and adds to Southern indebtedness. Such an observer might also note that from time to time, as in the United States' relations with Iran or Britain's with Iraq, while one part of the government is busy selling arms, another is preparing to wage war on the customer. Curiously, many Northern states which engage in this sort of chicanery are very touchy when their own citizens follow their example – the Mafia is, we believe, outlawed in several of them. Just how valuable this arms trade is to the North is illustrated in Table 10.6. As, in 1989 alone, expenditure on arms imports by developing countries was in excess of $100 billion it is obviously a sizeable factor, but the only passing reference to it in *Agenda 21* seems to be in Chapter 33, which is about financial resources for environment and development. In paragraph 18 (e) we find a tentative suggestion that we might just possibly raid some defence budgets for funds. Presumably this is a reference to the so-called 'peace dividend'.

As we might expect, the Global Forum was a little more forthcoming on both debt and arms spending. It produced documents on both issues and in them made the obvious links. Thus the preamble to the 'NGO Treaty on Militarism' opens with the words:

> In recognition of the links between militarism, debt, environmental degradation and maldevelopment, and in view of the fact that the UNCED process has thus far excluded these connections, we demand that the impact of militarism on the Earth, its people and on the global economy be put on the post-Rio Agenda.

Table 10.6
Military expenditure

	Expenditure % of GDP 1960	*1989*	*%age of combined education/ health expenditure 1990/88*	*Imports US$ millions 1989*	*Armed forces as % of teachers 1987*	*Armed forces per doctor 1987*
All developing countries	4.2	4.4	169	101,160	64	18
LDCs	2.1	4.1	146	11,770	108	77
Sub-Saharan Africa	0.7	3.2	108	6,220	89	76
Industrial countries	6.3	4.9	28	—	97	3

Source: extracted from UNDP, *Human Development Report, 1992,* Tables 20 and 41, pp. 166–7 and 200, Oxford University Press, New York and Oxford 1992.

Similarly the 'NGO Debt Treaty' spells out, in its ninth paragraph, the wider effects of Southern indebtedness, particularly when it results in enforced policies of structural adjustment:

> Such policies result from the transfer of sovereign decisions to the realm of creditors and interfere with the social, economic, commercial and technological policies of the Southern countries. Debt-for-nature swaps and buy-back mechanisms do not resolve the debt nor the environmental crisis and do not contribute to the development of policies consistent with the democratic management of resources.

Both 'treaties' express the problems simply and clearly, but neither offers much in the way of solutions beyond 'pledges' to work for the ending of various forms of injustice. Just how those pledges are to be redeemed is not yet clear, particularly in view of some of the structural difficulties faced by many Northern NGOs.

We have already pointed out that NGOs were excluded from any serious participation in the Earth Summit, but many of them published work beforehand in the hope of influencing delegates and of making major points to the world at large on the back of the considerable publicity which surrounded the event. *Caring for the Earth,*[7] which we have mentioned earlier, is one of the most important of these publications. It is important simply because of the weight wielded by its sponsors in the world of Northern NGOs and Northern state establishments. The main section dealing with debt comes in Chapter 9, 'Creating a Global Alliance', where it is suggested that official debt, that is what governments and para-statal agencies have lent, should simply be written off, but it goes on to suggest that commercial debt, borrowings from private banks, should be 'retired'. As part of the 'retirement' process 'debt-for-nature swaps' are proposed with safeguards against the taint of 'conditionality' (we have already discussed this particular nostrum).

No attention is paid to the degree to which the international banking system is increasingly dependent on massive interest payments from the developing world, nor to the improbability of such debt relief in the middle of a major recession. That banking system sometimes seems curiously fragile; once again the effect of the collapse of the publishing empire of Robert Maxwell on one of the North's major banks is instructive. A possible loss of US$2.75 billion resulting from the collapse has rocked the National Westminster Bank; the loss of interest and repayments from the South, which would be far greater than any loss caused by Maxwell, would probably call into question the viability of the entire Northern banking system. However, there is no immediate prospect of this. If Southern debt is to be eliminated, then we might expect to see the process beginning in the World Bank and the IMF, but Tables 1.3 and 1.4 show what is actually happening. The World Bank, a little more sensitive to criticism than the IMF, managed, until 1990, to conceal its depredations on Southern economies not by forgiving any debt, but by speeding up its disbursements, particularly in Africa. This meant that the Bank was able to show a net outflow of money until 1991. The IMF has simply gone on with business just as before and is engaged in making princely returns from its investments.

The World-Wide Fund for Nature and its fellows seem to be an improbable group of boat-rockers. Admirable as much of what they all do may be, it is difficult to see them pushing very hard for measures which could lead to grave difficulties for the institutional bastions of the North. Yet if the promises of the Global Forum are to be kept, then these powerful Northern NGOs are exactly the organisations which must be persuaded to take part in those campaigns which might well lead, if they are successful, to a radical restructuring of Northern banking and economic ways. WWF has been at the forefront of the debt-for-nature swaps and it may well take a substantial leap of its collective imagination to move away from that kind of protective conservationism. If it is to join that movement of NGOs begun at the Global Forum, then it, together with all the other powerful Northern organisations involved in development, must begin to adopt more fundamental and, we believe, ultimately more successful ways of sustaining a world in which people and all other diverse forms of life can co-exist. We shall watch their progress with interest.

11
Rolling Down From Rio

We have painted a very gloomy picture, but as the condition of the 'wretched of the earth' is substantially worse than it was thirty years ago, when hope rode a good deal higher than it does at present, we feel that it is warranted. Undignified claims by increasingly conservative governments, particularly in Britain, that 'we are not turning our back on the poor' are a fairly reliable indication that they are busy doing just that. As even the modest little half-promises made on behalf of the environment at Rio begin to recede into political history – the British government was among the leaders of those accepting, at the meeting of the IPCC in Copenhagen,[1] the proposal to postpone until 2030 the phasing out of the principal chemical threat to the ozone layer – and the developmental element of the conference is completely forgotten, we are entitled to the occasional qualm.

It has often been observed that a large part of the problem lies in a widespread and growing fundamentalist and fanatical religious sect. A strident and uncritical faith in the systems of Northern banks and a passionate belief in the divine mission of TNCs to generate profits by any means possible as the only ways to escape the fires of hell are the marks of the new zealots. Their theocracy has a secular arm in what they call, with an unexpectedly exquisite sense of irony, the 'free' market. With this institution, which is little more than a continually refined set of trading agreements designed to protect the gods Profitability and Laissez-faire, and which commands greater and greater authority, they coerce the unbeliever into abject submission. Ultra-orthodox fundamentalists spawned monetarism; in its pure form it proved to be heretical and is dying out, but it has left its cultural mark on the body of economic doctrine with dire consequences for the poor.

Dedicated and unregulated profit-hunting involves, among many other things, the use of cheap labour and of fairly disposable technologies which as well as killing surprising numbers of people also poisons and otherwise degrades the environment. Resources have to be secured and again, at whatever cost in lives, livelihoods, cultural and environmental damage, attached to manufacturing by the cheapest methods possible. This religious system creates castes of winners and losers – the latter must always, by definition, be a larger group than the former. Because the ecclesia is gimcrack, ramshackle, it sometimes collapses, some winners become losers, new winners are recruited and the believers are off again. For example, the older European arch-dioceses, including the Soviet schismatics, have been replaced as world powers by the imperium of the United States while Japan creates its own orthodox rite. But, as we all know, episcopacy corrupts and archiepiscopacy corrupts absolutely. Each new bench of these bishops seems a little more entropic than before.

Galbraith, in a lecture given to the Institute of Public Policy,[2] remarked that the developed industrial economies contain a division between 'those who pay and those who need' and that 'the modern mixed economy depends on the functional underclass'. In the last hundred years the continuous political battles waged by that under-class have successfully taken off 'the cruel edge' of capitalism and in doing so have effectively rescued the system from collapse. Minimum standards of education, health-care, pensions, welfare and fairly high employment, together with the regulation of the wilder excesses of trade, are the mark of the moderately civilised state. Mixed economies involving some state intervention and certainly some regulation may well be, as Galbraith remarked, the only system available to the contemporary world and it may be true that it is fruitless to search for another. In the sense that to look for some ultimate and overall solution to the world's economic ills can only be a function of utopian thinking, we may agree. But that minimal level of civilisation for the Northern under-class was fought for through a long and often bloody labour history and, as recent experience demonstrates, is always put in danger at those moments when any economy or group of economies is depressed. Capital, by its nature, will always seek to squeeze costs. The point is that the battles have yet to be fought by, and on behalf of, that much greater under-class in the South which, so far as it is developing at all, is doing so by gradually being drawn into the system. At present the South faces the full effect of the 'cruel edge' of capitalism. Those battles, as well as the internal contradictions of the system, so well mapped by Karl Marx and so dramatically illustrated by its current shakiness, will inevitably and at the very least modify the nature of liberal capitalism. But it does not follow that the modification will be any improvement on the *status quo ante.*

In the course of these crises of both faith and society UNCED was created and the question it is necessary to ask is whether it could be counted as a success. It would be foolish not to recognise that the very fact of the Summit was some sort of triumph of reason. It would be equally foolish to suppose that success in getting the rich and powerful to the water would persuade them to drink very much; those particular horses prefer gin. We have complained that what should have been a developmental conference was emasculated by an insistence that the most important issues were environmental. But to imagine that the United Nations could succeed in convening the largest summit meeting in history solely to examine ways of meeting the needs of the poor and redressing inequity was to ignore the realities of Northern social disintegration. The triumph consists in having persuaded powerful governments, for a moment, to see that the threat of humanly induced environmental catastrophe, exacerbated by poverty, was also a threat to their interests.

Politicians have a notoriously limited ability to pay attention to anything for very long and, as the globe has not got noticeably warmer in the last six months and seas have not risen, the environment has, for the time being, been displaced from their concerns by the great Northern slump. Nevertheless there is evidence that the public awareness which forced the creation of UNCED in the first place and which has begun to recognise the intimate connection between poverty and environmental destruction is still active. In a perverse sort of way, the recession even drives the point home as millions of unemployed Northerners watch their own environments crumbling for lack of investment. We may decently entertain a hope that the Rio conference marked a stage, not easily to be reversed, in the popular consciousness of the problems.

Those minimalist agreements should not completely be derided either. Two conventions, one on climate, the other on biodiversity, together with the General Statement of Principles on Forests, even though they are all seriously flawed, make up a surprisingly high score for an international summit meeting. The most important of them is the Convention on Climate and what it has done is to change the focus of the debate. While it committed those pusillanimous heads of state to almost nothing and, in any case, has to go through a long process of ratification before it is widely accepted, its existence at once acknowledges the gravity of the situation and may ultimately be used as a lever to compel further and better agreements.

We have also to bear in mind that cautious document, *Agenda 21*, and in particular Chapter 38, entitled 'International Institutional Arrangements':

> In order to ensure the effective follow-up of the Conference, as well as to enhance international cooperation and rationalize the intergovernmental decision-making capacity for the integration of environment and development issues and to examine the progress of the implementation of Agenda 21 ... a high-level Commission on Sustainable Development should be established in accordance with Article 68 of the Charter of the United Nations.(paragraph 11)

Such an organisation, if it proves to be effective, could go a long way to redressing the balance between developmental and environmental political agendas. Whether it will be is uncertain, simply because the provision was added that this 'Commission on Sustainable Development' should be responsible to the United Nation's Economic and Social Council (ECOSOC), a notoriously slow-moving, even slumbering, section of the UN. Should this turn out to be the case we need not lose all hope; whatever the balance of probability, it is after all possible that ECOSOC could be revived.

All these things, together with the strengthening of UNEP and UNDP, both of which moves arose from the conference, are gains and we must not ignore the long-term possibility of building on them. But exactly what can be built is less clear, if only because UNCED largely, as we have said repeatedly, avoided the structural causes of under-development, poverty and environmental degradation. We do not think that either the General Assembly of the United Nations or any of its subsidiary organisations will, of themselves, be able to engage any more closely with the problems of debt and trade than did the Earth Summit. Inevitably, because the UN is dominated by the industrial world it is compromised; while important issues must be raised in the General Assembly, and even more so in the fora provided by the UN's satellite organisations, they are unlikely to be settled there. No matter what support the United Nations is able to offer Southern economies and institutions, these issues have to be fought out with the banks, the TNCs, the major trading blocs and, in some instances, individual states.

We talk of fighting battles and have adopted Galbraith's description of unregulated capitalism: this language is confrontational, but it is a language generated by the system itself. Despite the dogmatisms of its extremist devotees it is a very flexible system, but it moves by continually resolving, or by failing to resolve, both its internal and external stresses. What we have to consider is how both to promote that movement and to ensure that it travels more or less in a civilising direction rather than along the savage path to which it naturally inclines. Times are bad, but they may not be hopeless and we must consider what there is that can be pressed into the service of compelling the advanced industrial economies to adapt to a more equitable world.

Self-interest will, of course, produce some change. Recent events have shown that slumps and widespread extremes of poverty destroy the natural markets both among the under and the middle classes. Some, even very large, corporations have collapsed as a result. Eventually, if it is to find ways out of the depression, private capital, aided by state intervention, will probably be compelled to invest, in one form or another, in the sorts of venture that will allow sufficient financial improvement among their customers to allow the market to grow again. If, or when, they do so there is a sporting chance that at least some of that investment will benefit deprived Southern areas. But change of that sort will be haphazard and limited by what is needed for capital to survive and it will always tend to concentrate on the industrialised parts of the world. Once again we may see that concentration taking place in the changing relations between the Eastern European economies and the West.

What followed, and may even be a result of, the collapse of the Soviet Union is not a more peaceful world able to make productive use of capital previously tied up in monstrous arms races, but increasing chaos. Former parts of the USSR and of Eastern Europe have collapsed into bloody conflict and racism, never far from European and American consciousness, has erupted throughout the Northern world, justified in the name of a revolutionary nationalism in which humanity is defined by blood. Capital, having successfully destroyed the greater part of Northern culture and replaced it with plastic-wrapped 'heritage', is engaged in a massive onslaught on its own poor as it also turn its big guns on the South. Hovering over this mess is a collection of very mobile, flexible and unaccountable TNCs, anxious only to preserve a sufficient purchasing base for their ranges of products. Neither the bunkered North nor the TNCs are seriously concerned with the very poor, wherever they may be, apart from neutralising them as a threat.

There are unlikely to be any swift resolutions of the debt crisis – morality counts for little in these confrontations and fanatics have always burnt their victims at the stake. At present debt-servicing is providing a useful cushion for a shaking and rather inefficient banking system. In the ordinary way we might hope that when (we must add 'and if') the North emerges from its slump and begins once more to invest on a wider scale, the few 'newly industrialised countries' will be added to and some of the larger agricultural economies will also pick up. If this were accompanied by an improbable new caution in the matter of lending and borrowing, or if, by some miracle, the banks recognised that it was counter-productive to engage in usurious lending, then debt might become a smaller factor at least among the 'middle-income' countries. There is, we suppose, always a small chance that in that event the banks, particularly the Tweedledum and Tweedledee of the World Bank and the IMF, will realise that the crippling and unmeetable debts of the very poor should gradually be forgiven, but there is little evidence in support of this pious hope.

Trade remains as one of the most intractable difficulties, because regulating its practices strikes at the very centre of the fondest beliefs of our free-market priesthood. Somehow the Clintons, Dunkels and MacSharrys of this world must be persuaded that their best interests are served by protecting the poor from the ravages of the NAFTA and the GATT. Their willingness to sacrifice small farmers and, incidentally, much of the environment, to an agreement in which only the destructive forms of agribusiness can survive suggests that this may be an uphill task. But even here there is yet hope; the Japanese have already begun to push hard to allow more processed goods from the South to be imported into Northern markets and early in the next century, when the MFA is finally ended, there may be some progress. But the greater part of another ten years will reduce an even larger part of the world's poor to absolute penury: improbably immediate change is needed. Businesses of one sort or another tend to have somewhat longer views than most governments, but they still look for what they are pleased to call 'economic returns' in the relatively short term. They may well realise that in the far longer run the reduction of massive worldwide inequity will increase markets, but their need for high rates of profit, usually calculated on three- to five-years spans and particularly during general slumps, confines them to the protectionist and extractive trading patterns that are a large part of the problem.

Debt and the terms of trade, together with rigged commodity pricing are not the only problems faced by the poor, but they are, so to say, the structural buttresses of poverty. Unless they are substantially modified, then matters can only get worse. There are many possible agents for change, quite aside from Northern self-interest. Southern states can frequently manoeuvre for better deals and, in some cases, introduce reforms which will make a difference to the plight of their poorest. But they must move cautiously; the lessons of countries like Guatemala, Nicaragua, Angola and Mozambique show what happens if they step too far out of line. Because the North is so insulated the increased destruction of Southern environments is unlikely to produce much more than humanitarian concern (although it would also be wrong to under-estimate the power of that). Environmental damage in the North will probably be dealt with piecemeal and only when matters become extreme, as in the case of the ozone layer, will some sort of international co-operation become a reality. Even then it is unlikely that the connection between poverty and environmental catastrophe will be made. We are left with what is probably the most powerful instrument of change, though it is one which still has a long way to go to become really effective. It is, on the one hand, the coalition between small, and often informal, grass-roots groups of activists throughout the developing world with their more organised fellows in Southern NGOs and, on the other, the increased working together of these groups with the international NGOs. In the end what is to be addressed is political and these are the makers of the new politics.

As UNCED so interestingly demonstrated, NGOs are very various, but in the North we are really only concerned with two kinds – those involved in developing countries and those concerned with the environment. Of the latter, two in particular make excellent examples: Friends of the Earth (FOE) and Greenpeace. They are both campaigning groups, but their methods differ. FOE, in addition to an extensive environmental monitoring programme, spends much of its resources and energies in educational and lobbying activities. Greenpeace, while it too engages in many of the same activities as FOE, is chiefly known for its campaigns of direct action. Both organisations have done much to raise a popular environmental consciousness; in some instances they have also succeeded in compelling change in both state and industrial practices. What neither of them has attacked in any serious sense is the nexus of state and commerce which leads to the practices they so ardently condemn. That they understand it is clear from the comments of Fiona Wier of FOE and Bill Hare of Greenpeace after the grubby proceedings at Copenhagen. But it is at least possible that their scope for going very far beyond campaigns for the conservation and protection of the environment is limited for many of the same reason that UNCED was limited – the environment is not the only place to begin and campaigns on its behalf must be made to run alongside the needs of the poor.

Among the NGO giants it is fascinating to watch the efforts of the World-Wide Fund for Nature to recognise that, contrary to the rumoured views of the Duke of Edinburgh, the poor are indeed 'an endangered species'. WWF is gradually changing its emphasis from simple conservation to considering ways in which people and the world's ecosystems, of which of course humankind is a part, can flourish together. This move could result in the organisation becoming a bridge between environmental concerns and developmental needs. Nonetheless WWF, more than any other NGO, fostered the popular conservatory perception of 'the wild' which, in turn, created the climate in which FOE and Greenpeace could grow. Between them they have sold the wild to a world anxious to believe that it still exists and which spends substantial sums of money to look at it through a camera lens. So it is that there is a real sense in which all three groups are, at present, essentially consumer organisations. While protecting the wild is no bad thing, making it a priority can easily lead to the humanly destructive political mistake typified by the debt-for-nature swaps.

In what could almost be seen as the final *cri-de-coeur* of the decent liberal ways of the older Northern NGOs, John Clark,[3] writing during a sabbatical from Oxfam, made a brave plea for 'people-centred' not 'dollar-centred' economics, for a more human world, for aid to the South which carries fewer conditions, for less destructive terms of trade and so on. No good NGO worker can resist coining an acronym, it is an occupational hazard: 'Good governance, then, calls for *Development, Economic Growth, Poverty alleviation,*

Equity, *Natural resource base preservation*, *Democracy* and *Social justice*. The DEPENDS formula for just development.'[4] Strained, perhaps, but it will do. NGOs are urged to push for this programme both in their policies in the South and in their advocacy everywhere.

Since Clark wrote, in 1990–1, the world has become a good deal grimmer and the camps more firmly entrenched. We do not doubt that he is right in his general view of what is needed, but both he and the NGOs concerned must face up to the obvious fact that their programme can no longer be seen as neutral humanitarianism – it is the very stuff of politics. Clark makes the common liberal mistake of assuming that this is to give in to a choice between the status quo and hoping for revolution. We suppose that the liberals' vision of the latter is something like the events in Ireland in 1916, or Russia in 1917, or Germany in 1933, entailing a cinematographic iconography of barricades, last stands and jack-boots. There are other models, even mixed ones like the Industrial Revolution – what Clark and his NGO comrades must grasp is that their aims in the present world situation, if achieved, will constitute a revolution. Any attempt to back off from this recognition will add to the not infrequent Southern view that they are merely collaborators in Northern hegemony.

The reasons for this are simple. As the North grows daily more xenophobic and as its economies stagger through new crises and as the GATT, the NAFTA, the nascent MTO and all the other trade barriers go up, so the space for the honest broker between rich and poor diminishes. Oxfam provoked a storm from the thought police by opposing apartheid. By opposing a whole range of financial, trade and racial boundaries, as it and its fellow groups must if they are either to retain credibility or to achieve their ends, they will be unable to maintain the pretence that they are outside politics. It is important to spell this out, because one of the mechanisms by which capital has maintained its control has been to persuade the world that it is the 'reasonable' centre and that serious, as opposed to cosmetic parliamentary, opposition is the dangerous and dirtily political fringe. Capital controls education and health to suit its own political ideologies and complains that criticism is 'politically motivated'. NGOs, if they are to be effective, must seize the *political* initiative, even if they risk losing state approval and thus their 'charitable' status.

In Chapter 9 we remarked that the natural allies of the international NGOs are not the governments of the world in which they began, but the Southern NGOs – above all those which have sprung from local needs and political battles. For obvious reasons the rural organisations of the developing world are what first spring to mind, but the alliances must be strengthened in urban areas too. There are plenty of examples of Southern NGOs which cross even that divide – the SEWA is only one – but even the most cursory examination reveals that many, possibly most, of them are deeply political and even, as in the case of women fighting for literacy in Latin America, 'revolutionary' groups. Field-workers in the international NGOs are frequently well aware of this and a high proportion of them are happy to work with essentially political demands in mind. Our view is that it is time that the Northern groups broke cover and mounted far more outspokenly political, rather than their usual alleviatory, campaigns. If they were to consider mutual collaboration and, dare we mention it, organisation, they might even succeed in frightening off the Charity Commissioners.

Southern NGOs do not need our approval, though they may welcome our help and it is here that the dangerous ground appears. Whatever help Northern NGOs, or Northern individuals, may offer their chief concern must, in the end, be to alleviate the pressures the North is bringing to bear on the South. This means that no matter what 'knowledge' NGOs may feel that they possess about the right ways in which to proceed, no matter what 'expertise' they are able to provide, the administration, use and finally the appropriation of that knowledge and expertise must be by the people of the South. We do not feel that it is very likely that this lesson will be completely absorbed in NGO corridors of power until their hierarchical and, indeed, patriarchal structures are replaced. Not all white, middle-class men are automatically politically incorrect and compromised, but organisations more or less entirely controlled by them almost always are.

The road from Rio is a rocky one and there is no guarantee that it will lead to a particularly salubrious end. Much of the world is caught up in economic and political transition in which the automatic response of all states is defensive. Capital itself is changing, finding a new equilibrium, much as the climate has in past aeons, and quite probably with results, at least for the poor, just as catastrophic. UNCED, with its predominantly environmentalist concerns, was a symptom of that change and may even have worked, to a degree, as a part of the Northern defences. However, it was also at Rio, in the Global Forum, that many NGOs came together for the first time and although their response in the form of the parallel 'treaties' was weak and confused, it was political, it did spell out the nature and the priorities of the campaigns to be fought in this generalised war. It may be that the increase in inter-organisational communication in the form of their electronic-mail 'conference' may become a vehicle for far more positive and effective action.

We do not yet know, either, how a stronger UNDP and the new Commission for Sustainable Development will affect matters; after all the Trojans pulled the horse into Ilion themselves. UNEP may be, as it were, another foal of that horse but the auguries are not good. Although it is environmental in its concerns, it has long paid lip-service to the importance of sustainable human policies as a guiding principle in environmental progress. Despite this its major contribution to the Earth Summit was to smuggle into the outcome that stale agenda of desertification. Thus its three most successful programmes, GEMS (the Global Environmental Management Survey), Regional Seas, and now that to do with desertification, are all programmes that do not bear much on the concerns of most people. Because UNCED was an opportunity not easily to be repeated, the flaccidity of its response to world crisis, typified by the importance it gave to this sad move by UNEP, seems paradoxically to have spurred many of those around it to greater effort, and if their room for manoeuvre is reduced it is not yet gone altogether. In a way, one interesting effect of the collapse of the Soviet Union, in spite of the chaos to which we have already alluded, is to have simplified political objectives even for those who work in such ambiguous organisations.

From time to time the ultimate sin against the Holy Ghost, that for which there can be no forgiveness, has been identified as final despair. We must try not to commit it. That great revolutionary ideologue, John Bunyan, calls it a 'miry slough' and says that in order to make the ground firm 'at least twenty thousand cartloads, yea, millions of wholesome Instructions, that have at all seasons been brought from all places' have been swallowed up by it. But Help, who pulls Christian out of the slough, seems to think that common sense will show the steps through.[5] Perhaps it can; but for the purposes of this commonwealth, common ground as well as common sense has yet to be found.

Notes and References

Introduction

1. *Irish Times*, 23 April 1993.
2. 'The philosophers have only interpreted the world, in various ways; the point is to change it.' 'Theses on Feuerbach XI', taken from Karl Marx, *Early Writings*, p. 423, Penguin Books, Harmondsworth 1975.
3. World Commission on Environment and Development (WCED), *Our Common Future*, Oxford University Press, Oxford 1987.
4. Ibid., pp. 43–4.
5. *Guardian*, London, 17 January 1992.
6. United Nations, *World Economic Survey 1991: Current Trends and Policies in the World Economy*, New York 1991.
7. John Pilger, quoting Studs Terkel, describes the process well in the introduction to his book *Distant Voices*, Vintage, London 1992.
8. *Guardian*, London, 21 December 1991.

Chapter 1

1. ICIDI, *North-South, A Programme for Survival*, London 1980.
2. ICIDI, *Common Crisis*, London 1983.
3. UNCED, *Agenda 21*, chapter 38, see paragraph 21 et seq.
4. WCED, *Our Common Future*, p. ix, Oxford University Press, Oxford 1987.
5. Ibid., p. 43.
6. Ibid., pp. 49–54.
7. IUCN, UNEP, WWF, *Caring for the Earth*, p. 21 and passim, Earthscan Publications, London 1991.
8. WCED, *Our Common Future*, p. 52.
9. Ibid.
10. Ibid., p. 95.
11. Ibid., pp. 122–3.
12. Ibid., p. 249.
13. Ibid., p. 78.
14. Ibid., pp. 314–23.
15. Ibid., pp. 80–3, 276.
16. Ibid., pp. 313–19.
17. Ibid., pp. 17–18.
18. UNCED, *Agenda 21*. See, for example, chapter 2, paragraph 24; chapter 33, paragraph 16, subsection (e).

19. WCED, *Our Common Future*, p. 81.
20. World Resources Institute (WRI), *World Resources 1992–93*, Oxford University Press, New York and Oxford 1992.
21. Hilaire Belloc, 'The Moral Alphabet', taken from H. Belloc, *Complete Verse*, Duckworth, London 1970.
22. Samuel Beckett, *Watt*, Olympia Press, Paris 1958.
23. United Nations, 'Revised Provisional List of Participants', New York, 8 May 1992.
24. WRI, *World Resources 1992–93*, p. 45.
25. Ben Ross Schneider, 'Brazil under Collor: Anatomy of a Crisis', *World Policy Journal*, vol. viii, no. 2 (spring 1991), p. 327. Quoted in WRI, *World Resources 1992–93*, p. 46.

Chapter 2

1. E. J. Hobsbawm, *Industry and Empire*, p. 13, Penguin Books, Harmondsworth 1968.
2. These remarks are based on the edited version of Margaret Thatcher's speech published in the *Guardian*, 9 November 1989.
3. Florentin Krause, Wilfred Bach and Jon Koomey, *Energy Policy in the Greenhouse: From Warming Fate to Warming Limit*, Earthscan Publications, London 1990.
4. Ibid., pp. 1.1–17; see also Martin Parry, *Climate Change and World Agriculture*, p. 12, Earthscan Publications, London 1990.
5. Krause et al., *Energy Policy in the Greenhouse*, p. 1.1–18.
6. Ibid., pp. 1.1–17–18.
7. Ibid., pp. 1.2–10–12.
8. Parry, *Climate Change*.
9. Ibid., p. 128.
10. UNCED, *Agenda 21*, para. 5.
11. WRI, *World Resources 1992–93*, p. 208, Oxford University Press, New York and Oxford 1992.
12. ETC report, *On the Road to a Sustainable Energy Economy*, p. 3, North Shields 1991.
13. United States Congress, Office of Technology Assessment, *Changing by Degrees: Steps to Reduce Greenhouse Gases*, Washington, DC 1991
14. World Bank, *World Development Report 1992: Development and the Environment*, p. 27, Oxford University Press, New York and Oxford 1992.
15. Karl Marx; see, for example, *Grundrisse, Notebook III*, p. 364 and passim, Penguin Books, Harmondsworth 1973.
16. WRI, *World Resources 1992–93*, p. 348.
17. Andy Crump, *Dictionary of Environment and Development*, p. 174, Earthscan Publications, London 1991; Parry, *Climate Change*, p. 10.
18. Inter-governmental Panel on Climate Change, *Scientific Assessment of Climate Change: Policy Makers Summary*, Geneva and Nairobi 1990.
19. Krause et al., *Energy Policy in the Greenhouse*, p. 1.3–1 *et seq*.
20. ETC, *Sustainable Energy Economy*, p. 36.
21. G. Marland et al., *Estimates of CO_2 Emissions from Fossil Fuel Burning and Cement Manufacturing Using the United Nations and the US Bureau of Mines Cement Manufacturing Data*, Oak Ridge, Tennessee, 1988, quoted in Krause et al., *Energy Policy in the Greenhouse*.
22. ETC, *Sustainable Energy Economy*, passim.
23. Ibid., p. 32.
24. IUCN, UNEP, WWF, *Caring for the Earth*, p. 1, Earthscan Publications, London 1991.
25. Ibid., chapter 10 passim.

Chapter 3

1. Henry Thoreau, *Walden*, 'Higher Laws', Boston 1854.
2. Commonly known as *Fingal's Cave.*
3. Charles Dickens, *Hard Times*, London 1854.
4. Roland Oliver and J. D. Fage, *A Short History of Africa*, Penguin Books, Harmondsworth 1962.
5. Don Hinrichsen, *Our Common Future: A Reader's Guide*, p. 30, Earthscan Publications, London 1987.
6. Ibid.
7. Robert and Christine Prescott-Allen, *Genes from the Wild*, p. 9, Earthscan Publications, London 1983.
8. Ibid., pp. 10–12.
9. Andy Crump, *Dictionary of Environment and Development*, pp. 169–70, Earthscan Publications, London 1991.
10. World Resources Institute, *World Resources 1992–93*, p.177, Oxford University Press, New York and Oxford 1992.
11. Ibid., p. 178.
12. Ben Wisner, *Power and Need in Africa*, pp. 48–9, Earthscan Publications, London 1988.
13. Edward B. Barbier, Joanne C. Burgess, Timothy M. Swanson, David W. Pearce, *Elephants, Economics and Ivory*, passim, Earthscan Publications, London 1990.
14. Preparatory Committee for UNCED, *Survey of Existing Agreements and Instruments, and Criteria for Evaluation*, Geneva 1991.
15. WRI, *World Resources 1992–93*, p. 309.
16. Ibid.
17. *Irish Times*, Dublin, 20 December 1991.
18. Preparatory Committee for UNCED, *Conservation of Biological Diversity, Options for Agenda 21*, Geneva 1991.

Chapter 4

1. Duncan Poore, *No Timber Without Trees*, p. 20, Earthscan Publications, London 1989.
2. Andy Crump, *Dictionary of Environment and Development*, p. 244, Earthscan Publications, London 1991.
3. *Guardian*, London, 1 November 1991.
4. Crump, *Dictionary*, p. 147.
5. K. Lai Chun, and Asmeen Khan, 'Mali as a Case Study of Forest Policy in the Sahel: Institutional Constraints on Social Forestry', ODI Social Forestry Network Paper 3e, London 1986.
6. Irene Dankelman and Joan Davidson, *Women and Environment in the Third World*, p. 69, Earthscan Publications, London 1988.
7. Binar Agarwal, *Cold Hearths and Barren Slopes: The Woodfuel Crisis in the Third World*, Zed Books, London 1986.
8. Dankelman and Davidson, *Women and Environment*, pp. 50–1.
9. Peter Walker, *Famine Early Warning Systems*, Earthscan Publications, London 1989.

Chapter 5

1. The authors acknowledge their debt in this chapter to the outstanding book on the GATT by Kevin Watkins entitled *Fixing the Rules: North–South Issues in International Trade and the GATT Uruguay Round*, CIIR, London 1992.

2. UNDP, *Human Development Report 1991*, p. 25, Oxford University Press, New York and Oxford 1991.
3. Liam Kennedy and Philip Ollerenshaw (eds), *An Economic History of Ulster 1820–1939*, p. 62 et seq., Manchester University Press, Manchester 1985.
4. Watkins, *Fixing the Rules*, p. 49.
5. WCED, *Our Common Future*, p. 85, Oxford University Press, Oxford 1987.
6. Ibid., pp. 53–4.
7. Walter Russell Mead, 'Bushism Found a Second Term Agenda Hidden in Trade Agreements', *Harper's Magazine*, pp. 37–45, New York, September 1992.
8. Ibid., p. 38.
9. Quoted in Watkins, *Fixing the Rules*, p. 134.
10. Ibid., pp. 128–9.

Chapter 6

1. WRI, *World Resources 1992–93*, p. 94, Oxford University Press, New York and Oxford 1992.
2. *Exodus*, chapters 8–14.
3. Reported in the *Guardian*, London, 3 October 1989.
4. Jaap Hardon and Walter de Boef, 'Appendix II: Background Paper', CPRO-DLO Centre for Genetic Resources (CGN), Wageningen May 1992.
5. E.G. Vallianatos, 'Agri-cultural Madness', in Hazel Waters (ed.) *Race and Class: The New Conquistadors*, pp. 89–106, Institute of Race Relations, London 1992.
6. Ibid., p. 91
7. Frances Moore Lappe and Joseph Collins, *World Hunger: 12 Myths*, p. 41 et seq., Earthscan Publications, London 1988.
8. David Weir, *The Bhopal Syndrome*, pp. 23–4, Earthscan Publications, London 1988.
9. P. O'Keefe and J. Kirkby, 'Mozambican Environmental Problems: Myths and Realities' in *Public Administration and Development*, vol. II, pp. 307–24, 1991.
10. Ibid., p. 322.
11. Quoted in Andy Crump, *Dictionary of Environment and Development*, p. 156, Earthscan Publications, London 1991.
12. Peter Walker, *Famine Early Warning Systems*, passim, Earthscan Publications, London 1989.
13. John le Carré, *Tinker, Tailor, Soldier, Spy*, chapter 34, Hodder and Stoughton, London 1974.
14. Irene Dankelman and Joan Davidson, *Women and the Environment in the Third World*, p. 9, Earthscan Publications, London 1988.
15. Marie Monimart, 'Women in the Fight against Desertification', in Sally Sontheimer (ed.), *Women and the Environment: A Reader*, pp. 40–8, Earthscan Publications, London 1991.
16. Nalini Singh, 'The Bankura Story: Rural Women Organize for Change', in Sontheimer (ed.), *Women and the Environment*, pp. 179–205.
17. WRI, *World Resources 1992–93*, pp. 94–6.
18. Kevin Watkins, *Fixing the Rules: North–South Issues in International Trade and the GATT Uruguay Round*, pp. 59–60, CIIR, London 1992.
19. Nigel Harris, *Of Bread and Guns: The World Economy in Crisis*, p. 210, Penguin Books, Harmondsworth 1983.
20. Ibid., p. 209.

21. Watkins, *Fixing the Rules*, p. 61.
22. Ibid., pp. 64–5.

Chapter 7

1. Don Hinrichsen, '*Our Common Seas: Coasts in Crisis*', p. 31, Earthscan Publications, London 1990.
2. *Independent on Sunday*, London, 1 November 1992.
3. Martin Ince, *The Rising Seas*, p. 59, Earthscan Publications, London 1990.
4. Hinrichsen, *Our Common Seas*, pp. 124–6.
5. WRI, *World Resources 1992–93*, Oxford University Press, New York and Oxford 1992.
6. A list of these may be found in *World Resources 1992–93*, p. 183.
7. Robin Clarke, *Water: The International Crisis*, pp. 92–105, Earthscan Publications, London 1991.
8. World Bank, *World Development Report 1992*, p. 197, Oxford University Press, New York and Oxford 1992.
9. Clarke, *Water*, pp. 8–12.
10. World Bank, *World Development Report 1992*, p. 48.
11. UNDP, *Human Development Report 1992*, Table 3, pp. 132–3, Oxford University Press, New York and Oxford 1992.
12. World Bank, *World Development Report 1992*, p. 106.
13. Ibid., pp. 107 and 109.
14. Clarke, *Water*, p. 12.
15. The Yarkon/Taninim aquifer, which lies beneath both pre-1967 Israel and the West Bank, provides between 25 and 40 per cent of Israel's water supplies. Israel is beginning to make use of it for the irrigation of parts of the Negev desert.
16. Czech Conroy and Miles Litvinoff, 'Dry-Season Gardening Projects, Niger', in their edited work, *The Greening of Aid: Sustainable Livelihoods in Practice*, pp. 69–73, Earthscan Publications, London 1988.
17. World Bank, *World Development Report 1992*, p. 46.
18. Clarke, *Water*, pp. 77–8.

Chapter 8

1. David Pearce, Anil Markandya, Edward B. Barbier, *Blueprint for a Green Economy*, passim, Earthscan Publications, London 1989. See also David Pearce, *Blueprint 2: Greening the World Economy*, passim, Earthscan Publications, London 1991
2. *Guardian*, London, 20 March 1992.
3. WCED, *Our Common Future*, pp. 226–7, Oxford University Press, Oxford 1987.
4. Figure quoted in Don Hinrichsen, *Our Common Future: A Reader's Guide*, p. 28, Earthscan Publications, London 1988. Taken from *Environmental Data Compendium 1986*, OECD, Paris.
5. WRI, *World Resources 1992–93*, Oxford University Press, New York and Oxford 1992.
6. Andy Crump, *Dictionary of Environment and Development*, p. 176, Earthscan Publications, London 1991.
7. Ibid., p. 186.
8. E.B. Uvarov, D.R. Chapman, Alan Isaacs, *The Penguin Dictionary of Science*, passim, Penguin Books, Harmondsworth 1979.
9. WRI, *World Resources 1992–93*, p. 144.

10. Hilaire Belloc, 'Matilda' in *CautionaryTales for Children*, London 1930. Quoted by permission of the Peters Fraser and Dunlop Group Ltd.
11. Quoted in *Geofile*, September 1987, no. 96, London.
12. WRI, *World Resources 1992–93*, p. 25.
13. UNDP, *Human Development Report 1991*, p. 20, Oxford University Press, New York and Oxford 1991.
14. UNDP, *Human Development Report 1991*, p. 134, Oxford University Press, New York and Oxford 1991.
15. Jorge E. Hardoy and David Satterthwaite, *Squatter Citizen*, pp. 197–8, Earthscan Publications, London 1989; David Weir, *The Bhopal Syndrome: Where Will It Happen Next?*, pp. 85 and 193, Earthscan Publications, London 1988.
16. A far more extensive list of major industrial accidents, from which these examples were taken, is to be found in Weir, *The Bhopal Syndrome*, pp. 187–95.
17. Ibid., p. 196.
18. *Chemistry and Industry*, 15 November 1980, quoted in Robert Allen and Tara Jones, *Guests of the Nation: The People of Ireland versus the Multinationals*, Earthscan Publications, London 1990.
19. Robert Allen and Tara Jones, *Guests of the Nation: The People of Ireland versus the Multinationals*, passim, Earthscan Publications, London 1990.
20. Preparatory Committee for UNCED, *Environmentally Sound Management of Toxic Chemicals*, Geneva 1991.

Chapter 9

1. Irene Dankelman and Joan Davidson, *Women and Environment in the Third World*, p. 21, Earthscan Publications, London 1991.
2. Frances Dennis and Dulce Castleton, 'The Guarari Housing Project, Costa Rica', in Sally Sontheimer (ed.), *Women and the Environment: A Reader*, pp. 147–62, Earthscan Publications, London 1988.
3. Paulo Friere was born in Recife, Brazil, in 1921. He has been a lifelong revolutionary thinker and activist who has concentrated his efforts on the means of restoring and validating the languages of the poor. He has seen this as an essential step by the poor in resisting the exploitative and marginalising pressure of the ruling and middle classes. His most famous book is *Pedagogy of the Oppressed*, Penguin Books, Harmondsworth 1969.
4. David Archer and Patrick Costello, *Literacy and Power*, pp. 122–3, Earthscan Publications, London 1990.
5. E.P. Thompson, *The Making of the English Working Class*, passim, rev. edn, Penguin Books, Harmondsworth 1980.
6. Dankelman and Davidson, *Women and Environment in the Third World*, pp. 35–6 (Burkina Faso), 37–8 (Kenya), 49 (Uttar Pradesh).
7. Archer and Costello, *Literacy and Power*, pp. 166–71.
8. Mrs C.F. Alexander, 'All Things Bright and Beautiful' (this verse is now usually excluded).
9. See Thompson, *The English Working Class*, passim.
10. The Barabaig are a pastoral people in East Africa. Their ancient right to graze cattle was severely compromised by CIDA's attempts to initiate new settled farming projects for other, land-hungry and displaced people within Barabaig territory. CIDA strenuously resisted any recognition of the pastoralists' rights. (Private communication from the author of an unpublished paper.)
11. John Clark, *Democratizing Development*, pp. 50–1, Earthscan Publications, London 1991.

12. See inter alia, Teresa Hayter, *Exploited Earth*, passim, Earthscan Publications, London 1989, and *Aid: Rhetoric and Reality*, passim, Pluto Press, London 1985; Susan George, *A Fate Worse then Debt*, passim, Penguin Books, Harmondsworth 1988.
13. WCED, *Our Common Future*, p. 328, Oxford University Press, Oxford 1987.

Chapter 10

1. World Bank, *World Development Report 1992*, p. 29, Oxford University Press, New York and Oxford 1992..
2. Quoted in WRI, *World Resources 1992–93*, p. 84, Oxford University Press, New York and Oxford 1992.
3. UNDP, *Human Development Report 1992*, pp. 3–5, Oxford University Press, New York and Oxford 1992.
4. Susan George, *How the Other Half Dies*, Penguin Books, Harmondsworth 1986, *A Fate Worse than Debt*, Penguin Books, Harmondsworth 1988.
5. Quoted from a discussion paper at the third preparatory committee for UNCED.
6. Franz Kafka, *The Castle*, chapter 15 'Petitions', from *The Complete Novels*, p. 395, Minerva, London 1992.
7. IUCN, UNEP, WWF, *Caring for the Earth*, Earthscan Publicaions, London 1991.

Chapter11

1. See Chapter 2 for details.
2. J.K. Galbraith, delivered in the House of Commons, 24 November 1992, reported in the *Guardian*, 25 November 1992.
3. John Clark, *Democratizing Development*, Earthscan Publications, London 1991.
4. Ibid., p. 245.
5. John Bunyan, *The Pilgrim's Progress*, part I, Nathaniel Ponder, London 1680.

Index